D0084734

# Ecology and Natural History of Desert Lizards

Princeton University Press
Princeton, New Jersey

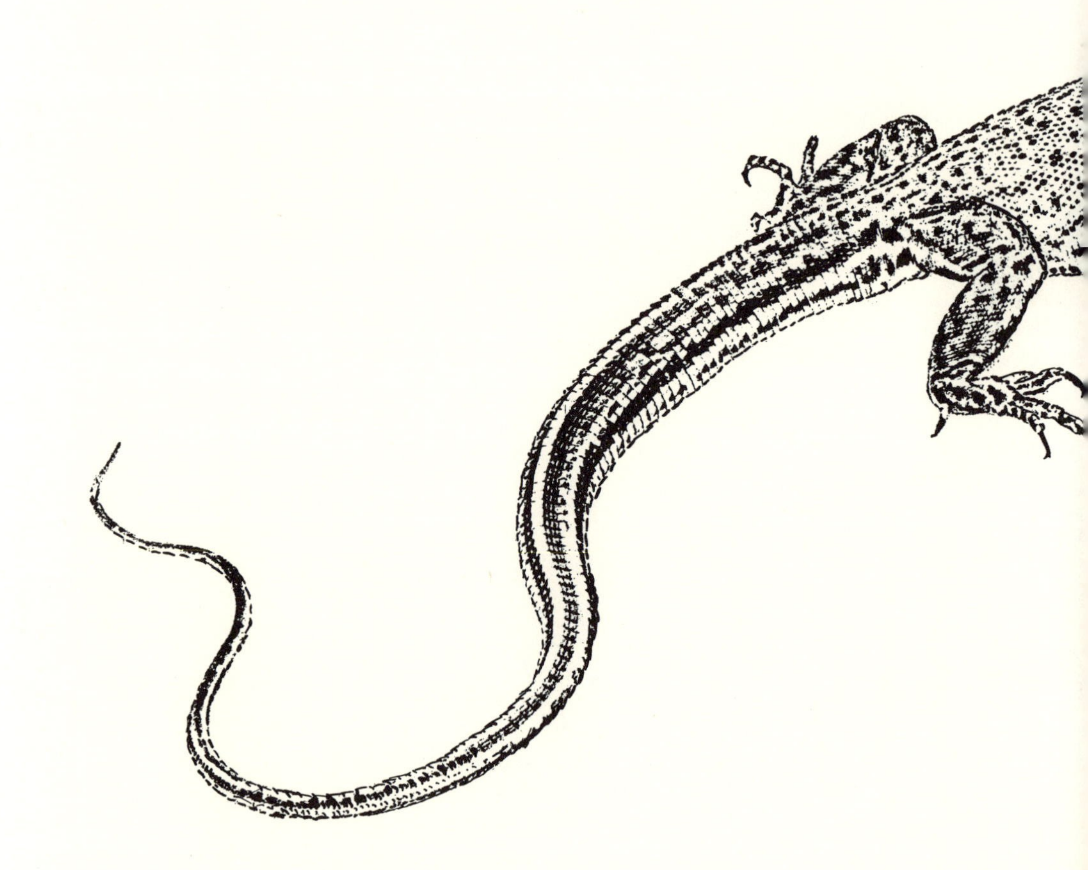

# Ecology and Natural History of Desert Lizards

## Analyses of the Ecological Niche and Community Structure

BY ERIC R. PIANKA

Published by Princeton University Press, 41 William Street
Princeton, New Jersey 08540
In the United Kingdom: Princeton University Press, Guildford, Surrey



Library of Congress Cataloging in Publication Data will
be found on the last printed page of this book

ISBN 0–691–0–08148–4 (cloth)
ISBN 0–691–08406–8 (pbk)

This book has been composed in Linotron Sabon

Clothbound editions of Princeton University Press books are
printed on acid-free paper, and binding materials
are chosen for strength and durability

Printed in the United States of America
by Princeton University Press, Princeton, New Jersey

*To Betsy, Matilda, Molly,*
*Brunhilda, Bertha,*
*and Old Lady*

# Contents

# Preface and Acknowledgments

For the past two decades I have been engaged in comparative research on the ecology of desert lizards. Much of this work has been published piecemeal along the way. In this monograph I develop an overview and undertake a synthesis. Old data and nomenclature are updated, some errors are corrected, new information is presented, while results reported earlier are often reinterpreted and conclusions extended and/or modified. Some redundancy with published papers has been inevitable: nevertheless, even a completely repetitive compendium could conceivably be of some utility in consolidating a widely scattered body of work under a single cover. Another function of the present summary is to make accessible considerable previously unpublished data, which might well prove useful to others in testing various hypotheses I have not even considered here.

Many people contributed to this effort, both in the field and in the laboratory. For field assistance and companionship I thank Carolyn Cavalier, Duncan Christie, Larry Coons, Helen Dunlap-Pianka, Rand Dybdahl, Bill Giles, Ray Huey, Paula Levin Mitchell, Hortense McGillicutty, François Odendaal, Ted Papenfuss, Nicholas Pianka, Bill Shaneyfelt, Michael Thomas, and Mary Willson. I am greatly indebted to Virginia Johnson Denniston and Glennis Kaufman for much careful help with data accession and analysis. A virtual plethora of work-study students, who performed a multitude of tedious tasks, must unfortunately remain unnamed, lest one be overlooked inadvertently. Michael Egan and Thomas Schultz painstakingly examined and identified the stomach contents of many thousands of lizards. I am indebted to Vicki Brown, Cathy Debus, Bill Giles, Marc DesParois, Janet Young, and Ann Winger for artwork. H. G. Cogger kindly made available a superb photograph and artistically augmented tone-line dropout rendition of *Varanus eremius*, upon which one line drawing is based. Glen Storr, A. R. Main and Wulf Haacke helped immeasurably in all sorts of ways, especially with bureaucratic red tape in Australia and South Africa. I am inexpressibly grateful to Larry Lawlor for allowing me to exploit his elegant and extensive software for community analyses. J. A. Joern shared in the agony and joy of developing subroutines for calculating statistics from multiple randomization runs. For reading various preliminary drafts of the manuscript and offering valuable suggestions for its improvement, I thank Ted Case, Ray Huey, Michael Kane, Robert Mason, Tom Schoener, and Joan Whittier. Colleagues Henry Horn, Ray Huey, Larry Lawlor, Robert MacArthur, Gordon Orians, William Parker, and Joseph Schall contributed ideas, friendship, and inspiration. For financial support totalling nearly one-quarter of a million dollars, I gratefully acknowledge the University of Texas Research Institute, the Los Angeles County Museum of Natural History Foundation, the National Institutes of Health, the National Science Foundation, the National Geographic Society, and the Guggenheim Memorial Foundation. Without this generous funding and extensive manpower, my perhaps overly ambitious research program could never have been undertaken, let alone be carried as far as I have been privileged to take it.

ERIC R. PIANKA
30° 14′ 43″ N. × 98° 14′ 45″ W.

# Notes on the Line Drawings

CHAPTER 1: *Chameleo dilepis*, a bizarre and uncommon arboreal lizard in the eastern Kalahari.

CHAPTER 2: The zebra-tailed lizard, *Callisaurus draconoides*, an American iguanid. These denizens of the open spaces make long zigzag runs when pursued.

CHAPTER 3: The "mountain devil," *Moloch horridus*, an Australian agamid, that specializes on ants as prey.

CHAPTER 4: The Kalahari lacertid *Nucras tessellata* is active during the heat of the day and is a dietary specialist on scorpions.

CHAPTER 5: Head study of a North American horned lizard, *Phrynosoma*, an ant-specialized iguanid ecologically equivalent to the Australian agamid *Moloch horridus* (latter appears with Chapter 3).

CHAPTER 6: *Gemmatophora longirostris*, an arboreal Australian agamid found on sandridge sites in the Great Victoria desert.

CHAPTER 7: Head study of an Australian varanid.

CHAPTER 8: The spinifex gecko, *Diplodactylus elderi*, a nocturnal denizen of *Triodia* grass tussocks.

CHAPTER 9: *Chameleo dilepis*, a bizzare and uncommon arboreal lizard in the eastern Kalahari.

CHAPTER 10: *Diplodactylus conspicillatus*, a terrestrial Australian gecko that is nocturnal and specializes on termites for food.

CHAPTER 11: The desert iguana, *Dipsosaurus dorsalis*, an American iguanid with an adult diet consisting largely of plant materials, including flower blossoms of the creosote bush, *Larrea divaricata*.

CHAPTER 12: *Varanus eremius*, a terrestrial pygmy monitor lizard that specializes on other Australian desert lizards for prey.

# Ecology and Natural History of Desert Lizards

# I  Three Independently Evolved Desert-Lizard Systems

Red sandridge areas in the Great Victoria desert of Western Australia support as many as 42 different species of lizards. In addition to a rich fauna of fairly typical lizards, these include small, nearly legless subterranean skinks and snake-like pygopodids as well as very intelligent large mammal-like varanid lizards. In terms of number of species, this region very probably sustains the most diverse lizard communities on Earth. Areas in the Kalahari semidesert of southern Africa, located at similar latitudes and practically identical in their basic climatology and general physical physiognomy, support about half as many species of lizards. Flatland areas in the warm deserts of the North American southwest are still more impoverished, accommodating only from 6 to 11 species of lizards, depending upon the complexity of their vegetation. Intercontinental comparisons provide some valuable insights into how these natural communities are put together.

Lizards constitute an extremely conspicuous element of the vertebrate faunas of most deserts, especially warmer ones. Indeed, the mammalogist Finlayson (1943) referred to the vast interior deserts of Australia as "a land of lizards." Like other ectotherms, lizards obtain their body heat solely from the external environment, as opposed to endotherms such as birds and mammals, which can produce their own heat internally by means of oxidative metabolism. Moreover, along with other ectotherms (Pough 1980), lizards are low-energy animals. Bennett and Nagy (1977) underscore the great "economy of the saurian mode of life" by pointing out that one day's food supply for a small bird will last a lizard of the same body size more than a month. Ectothermy presumably has distinct advantages over endothermy under the harsh and unpredictable conditions that prevail in deserts (Schall and Pianka 1978): by means of this thermal tactic (Chapter

3), lizards can conserve water and energy by becoming inactive during the heat of midday, during resource shortages, or whenever difficult physical conditions occur (as during heat waves or droughts). Birds and mammals must weather out these inhospitable periods at a substantially higher metabolic cost. Ectothermy thus confers lizards with the ability to capitalize on scant and unpredictable food supplies and other resources; presumably this gives lizards a competitive advantage over endotherms in many desert environments.

Most lizards are insectivorous and hence relatively high in the trophic structure of their communities: as a result of this and the fact that lizards attain such high abundances in deserts, they probably often experience relatively keen competition. Lizards usually take their prey items intact, greatly facilitating analyses of their trophic relationships. Indeed, lizards are exceedingly tractable subjects for the study of natural history and ecology (Huey et al. 1983). As Schoener (1977) put it, "lizards have not only proven ecologically exciting in their own right, but they may well become paradigmatic for ecology as a whole." Moreover, Schoener noted that "lizards are, in general, nearer to the 'modal' animal than birds; they are terrestrial, quadrupedal, poikilothermic, relatively slow-growing, and lack parental care."

With the help of a small army of colleagues and assistants over the past two decades, I have gathered and analyzed extensive data on the ecological relationships of saurofaunas of some thirty-odd desert study sites, which lie at roughly similar latitudes on three continents: western North America, southern Africa, and Western Australia. A series of representative flatland desert areas were selected for investigation on each continent; these study sites are homogeneous and continuous, but extensive enough[1] to facilitate sampling and generally well suited for ecological analyses. Lizards have evolved in response to desert conditions independently within each of these three continental desert systems. Intercontinental comparisons should reflect the extent to which the interaction between the lizard body plan and the desert environment is determinate and predictable. If observed, any convergence between such independently evolved ecological systems could conceivably provide insights into the operation of natural selection and might eventually help to lead to general principles of community organization.

[1] Study areas vary in size from about half a square kilometer to several square kilometers. Study sites are most widely spaced in North America and closest together in the Kalahari. Average distances to the nearest other study area for North America, the Kalahari, and Australia, in that order, are 190, 61, and 87 km., respectively (standard deviations 63, 31, and 120).

To gather data on the ecological relationships of these lizard faunas, my assistants and I walked thousands of kilometers through study sites observing lizards. We spent five full years in the field between 1962 and 1979 and nearly twelve man-years collecting data on lizards (sites were visited repeatedly over essentially the full seasonal period of lizard activity). Snakes, birds, and mammals observed on the study areas were recorded, and, in some cases, collected. Microhabitat and time of activity were recorded for most lizards encountered active above ground. Whenever possible, lizards were collected[2] so that their stomach contents and reproductive condition could be assessed later in the laboratory. Resulting collections of some 4,000 North American specimens, more than 5,000 Kalahari animals, and nearly 6,000 Australian ones—representing some 90-odd species in eleven of the fifteen extant families of lizards—are lodged in the Los Angeles County Museum of Natural History, the Museum of Vertebrate Zoology of the University of California at Berkeley, and the Western Australian Museum in Perth. For some purposes, data are augmented with other museum specimens and literature records.

In western North America (Figure 1.1), deserts form a continuous and enclosed series, uninterrupted by major physical barriers, over a latitudinal range of a thousand kilometers, from eastern Oregon and southern Idaho through Sonora and Baja California (Mexico). Three more or less distinct regions are generally recognized within the region: a northern Great Basin desert, a southern Sonoran desert (Shreve 1942), and an intermediate area in southern California and southern Nevada known as the Mojave desert. A fourth "Colorado desert" region in the area between the Mojave and Sonoran deserts is sometimes distinguished as well. Boundaries of the deserts and those between the various subdesert systems are not necessarily sharp. The Great Basin is a very simple desert structurally, with predominantly a single perennial plant life form (low "microphyllous semishrubs" including *Atriplex confertifolia* and *Artemesia tridentata*). Vertical heterogeneity is minimal. Because these small shrubs in northern deserts are usually very densely packed and uniformly spaced, horizontal spa-

[2] Nocturnal species were usually captured by hand after sighting them either by eye shine or body shine using Winchester head lights or a kerosene lantern. Diurnal species were collected in a wide variety of different ways, including shooting them with BB guns, .22 caliber "bird shot" or "dust shot," or a .410 gauge shotgun. A few lizards were captured in pitfall traps and many more were exhumed from their burrows by digging. Certain wary species, particularly *Varanus*, required tracking down of the animal. Some lizards were collected by burning *Triodia* clumps; Australian *Ctenotus* skinks often had to be collected by flattening spinifex tussocks with shovels.

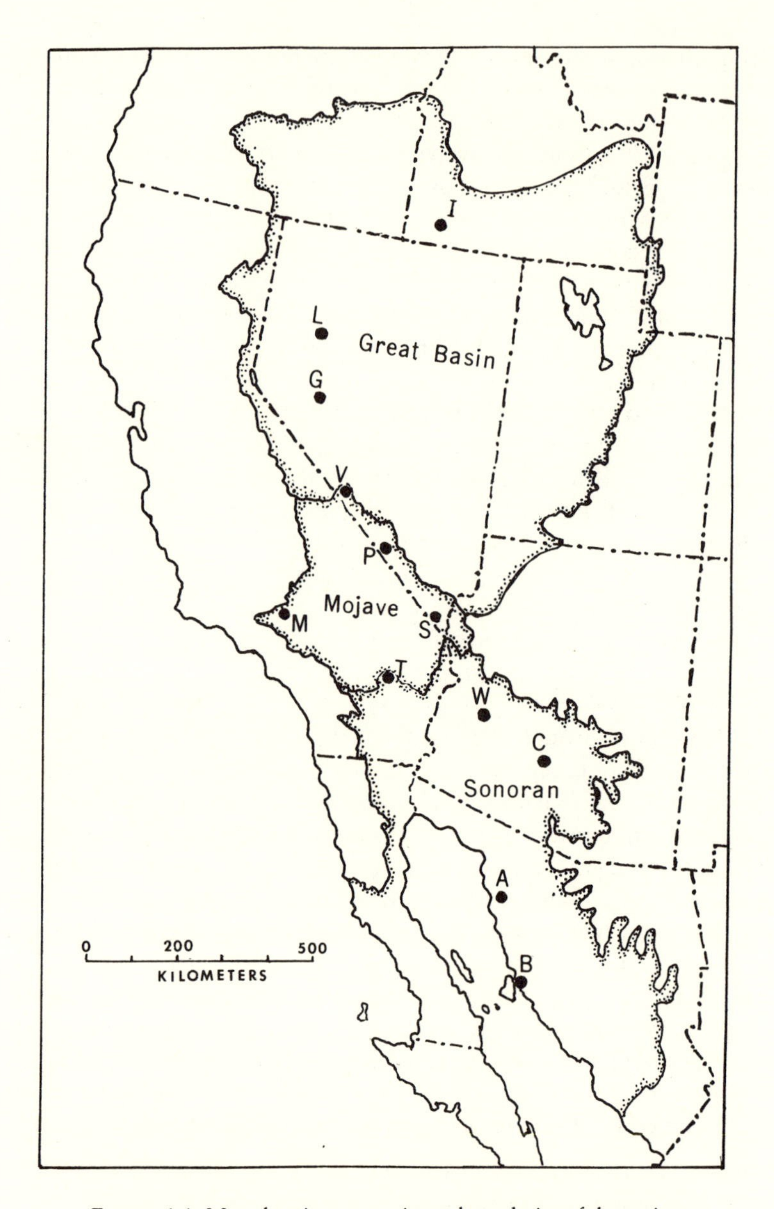


FIGURE 1.1. Map showing approximate boundaries of the various deserts in western North America (excluding the Chihuahuan), based on Shreve (1942). Positions of study sites are shown with identifying letter code names. For more exact positions, see Pianka (1967).

tial heterogeneity is also low. The Great Basin desert thus presents a very repetitive and rather monotonous landscape. In contrast, the Sonoran is an arboreal desert with several species of trees and, in addition to small semi-shrubs such as *Atriplex*, many large woody shrubs (the most conspicuous of which is creosote bush, *Larrea divaricata*). As a result, the vertical component of spatial heterogeneity is considerably greater in the Sonoran than it is in the Great Basin desert. The Mojave desert is also dominated by *Larrea* but is virtually treeless, except for a few *Yucca brevifolia* "trees." The diversity of different plant life forms reaches an apex in the Sonoran desert (Shreve 1951). The addition of perennial plant life forms in the southern deserts is often accompanied by fewer plant individuals per unit area and much greater horizontal heterogeneity in their distribution in space. Rivulets and dry washes along alluvial outwash fans (*bajadas*) are lined with different species of plants than intervening areas. Saurofaunas in the flatland deserts of western North America consist of a basic "core" set of four ubiquitous species (the whiptail *Cnemidophorus tigris*, the side-blotched lizard *Uta stansburiana*, the leopard lizard *Crotaphytus wislizeni*, and the desert horned lizard *Phrynosoma platyrhinos*) to which various combinations of other species are added, with the total number of species increasing from 4 or 5 in the north to 9–11 in the south (Pianka 1967). Three Great Basin areas were studied (areas I, L, and G), and four sites were selected in the Mojave desert (areas V, S, P, and M), along with a further five Sonoran sites (areas T, W, C, A, and B). North American field work was undertaken primarily during 1962 through 1964, but a few observations were also made during the summers of 1966 and 1969.

In the southern Kalahari (Figure 1.2), major physiognomic and vegetational changes take place along an east-west precipitation gradient: the more mesic eastern region consists of flat sandplains with a savanna-like vegetation, whereas stabilized sandridges characterize the drier western "sandveld" or "duneveld." These red sandridges, which average about 10 meters in height, generally parallel the direction of the strongest dominant prevailing winds, which blow in August and September. Sandridges are frequently as long as one kilometer or even more and support a characteristic grassy dune vegetation. Interdunal flats or "streets" average about 250 meters in width but may occasionally be much more extensive, sometimes as wide as several kilometers, with a vegetation consisting of various grasses, laced with large bushes and scattered small trees. Two common woody shrubs of Kalahari flats are *Rhigozum trichotomum* and *Grewia flava*, both of which are vaguely reminiscent of the North American creosote bush

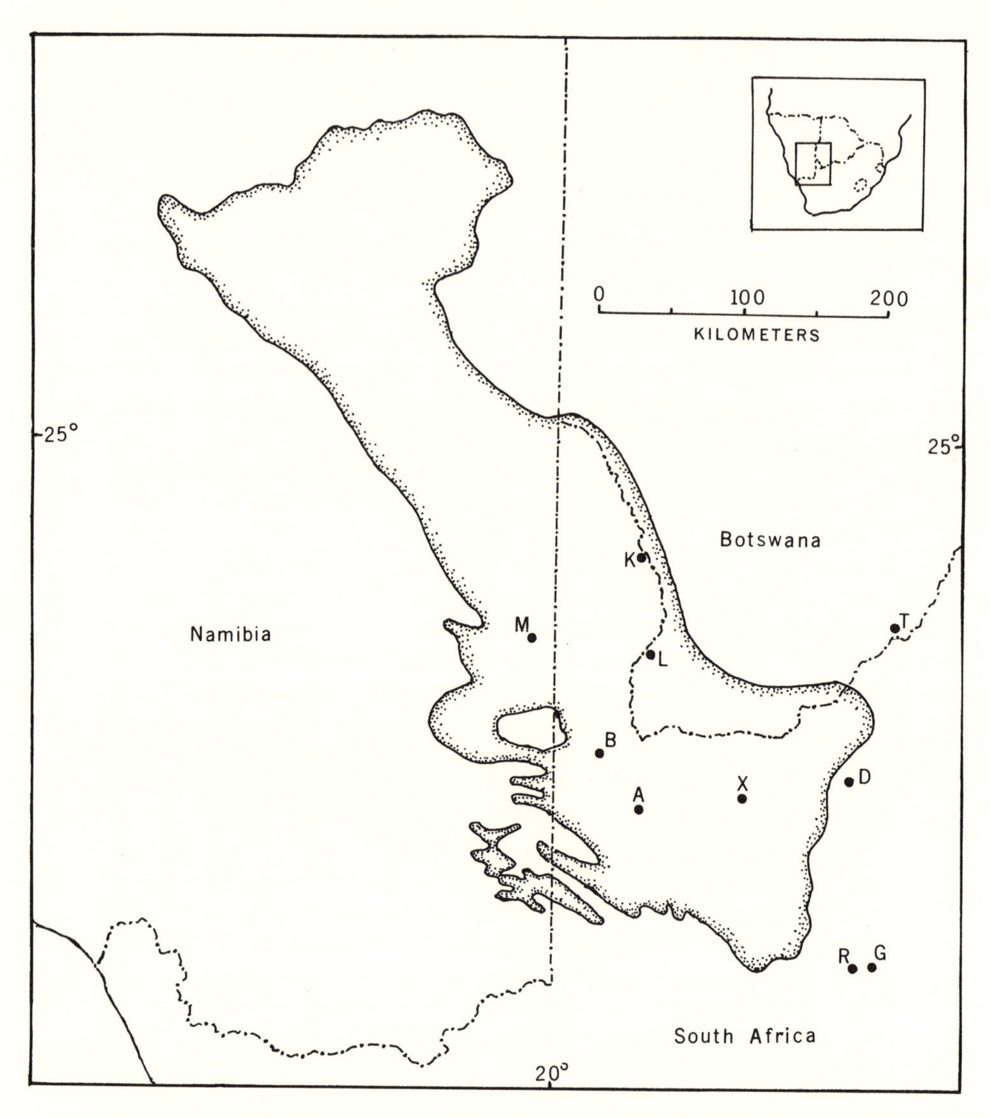


FIGURE 1.2. Map showing approximate boundary of the sandveld portion of the Kalahari semidesert in southern Africa. Sandplains, also considered Kalahari desert, extend beyond the sandveld, especially toward the east. This map is based on Leistner (1967). Study areas are shown along with identifying letter code names. For more precise locations, see Pianka (1971e).

*Larrea divaricata.* Detailed descriptions of Kalahari vegetation, with photographs, are provided by Leistner (1967). Ten study areas were selected representing the full range of habitats and conditions prevailing across the southern part of the Kalahari (Pianka 1971e; Pianka and Huey 1971). Average annual precipitation on four easternmost sites (designated areas G, D, R, and T) is greater than 200 millimeters, whereas annual rainfall is less than 200 millimeters on six more westerly study sites (areas L, K, M, B, A, and X). The latter six sites all lie within the "dune area" of the southern Kalahari as delineated by Leistner (1967), and all have sandridges or sand dunes. Areas L and K are within the Kalahari-Gemsbok National Park and thus represent a relatively "pristine" Kalahari desert habitat (both areas support lion, leopard, and hyaena, as well as a half dozen different species of antelope, for example). The four eastern sites are all on fairly flat terrain but vary in their vegetation: area G is a chenopod shrub desert with *Atriplex semibaccata* and area R is a nearly pure *Rhigozum* flat, whereas area D supports a more diverse mixture of small to large shrubs, including *Rhigozum, Grewia,* and the thorny bush *Acacia mellifera.* Area T, in southern Botswana, is a mixed, open forest and savanna site with a substantial number of trees (this site was still quite wild with Spotted Hyaena at the time of our initial study in 1969–70, but with the advent of bore water it has since become cattle-grazing country); the savanna and forest sections of area T can be treated separately as subareas. Most Kalahari observations were made during the 1969–70 Austral season and supplemented on a second trip in 1975–76.

The Great Victoria desert (Figure 1.3) is also predominantly sandy with red sands, and it supports a vegetation consisting mainly of so-called "spinifex" or "porcupine" grasses (genus *Triodia*) plus various species of gum trees (*Eucalyptus*). In wetter places and on harder soils, tracts of "mulga" (*Acacia aneura*) occur. Occasional dry lakebeds are inhabited largely by various shrubby chenopod species (including *Atriplex lindleyi*). Stabilized long red sandridges, parallel to dominant September-October winds and generally very much like those of the Kalahari, are scattered throughout the Great Victoria desert, particularly in the eastern interior. Extensive areas of sandplain occur as well. The region is very heterogeneous, and mixed habitats of shrubs, *Triodia, Acacia,* and *Eucalyptus* occur on desert loams (Beard 1974 describes and illustrates the vegetation of the region). Study sites A and M are on such ecotonally mixed areas. "Red Sands" (area R) and area E are sandridge sites with marble gum trees (*Eucalyptus gongylocarpa*), a few desert bloodwood trees (*Eucalyptus* sp.), extensive

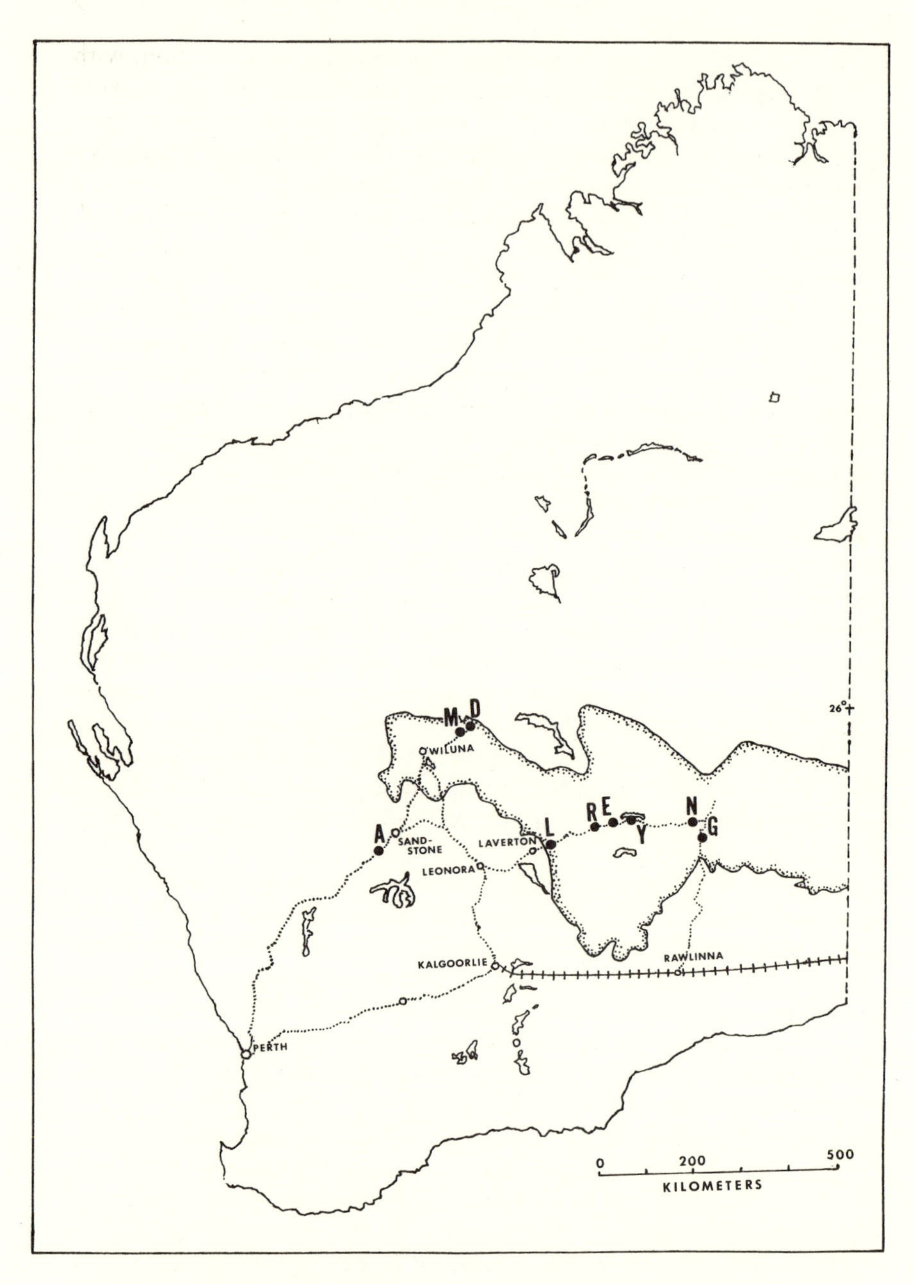


FIGURE 1.3. Map showing the approximate extent of the Great Victoria desert in Western Australia, based on Beard (1974). Study sites are shown with their identifying letter codes. For more exact positions, see Pianka (1969b and 1982).

spinifex (*Triodia*), and a variety of sandridge perennials. On area D, a few scattered smaller sandridges and sand dunes support a lower and more open vegetation with many fewer trees. Sandplain habitats also occur on these three study sites between sandridges. Two other areas, G and L, consist solely of such flat or gently rolling sandplains, with large marble gum eucalypt trees, spinifex, and some scattered bushes. Area N is a "pure" spinifex flat (treeless grass desert), whereas area Y is a relatively pure (nearly treeless) shrub desert site in a dry lakebed. The last site was chosen because the structure of its vegetation was essentially comparable to that of North American Great Basin desert areas as well as that of Kalahari area G (see also below). Two separate expeditions were undertaken to the Western Australian deserts, the first in 1966–68 when eight areas were selected for study (see Pianka 1969a), and the second in 1978–79 when two sites were chosen for more intense analyses (one of these, area L, was also among the original eight and thus was examined on both trips; see also Chapter 10).

As indicated earlier, these study areas vary considerably in numbers of lizard species. North American flatland sites support from as few as 4 to 10 species, whereas those of the Kalahari are somewhat richer, with between 11 and 17 species. In the Great Victoria desert, from 15 to as many as 42 species of lizards coexist in sympatry on a given site.[3] Censuses of the species of lizards occurring on the various study areas, as well as their approximate relative abundances, are given in Appendix A. The numbers of species in different lizard families that coexist on these study areas are listed in Table 1.1.

Eleven of the 15 recognized families of extant lizards occur in these systems. Figure 1.4 depicts a crude "cladogram" of their probable phylogenetic relationships, based on Estes (1983). Higher taxonomic levels contribute little or nothing to differences in diversity, since exactly five lizard families are represented in each continental desert-lizard system. At the generic level, Australian deserts are somewhat richer (about 22–25 genera, depending on whose classification is followed) than either the Kalahari (13 genera) or the North American deserts (12 genera). Although it may be somewhat confusing at first, I have followed the nomenclature of Storr, Smith, and Johnstone (1981, 1983, 1985), with the hope that Australian generic and species names will eventually be standardized.

[3] The sequence of my studies on these three continents was fortuitously quite fortunate: had I visited the Australian deserts first, I might well have been spoiled by the high diversity and would probably have found the North American desert-lizard system too boring and too redundant to complete an adequate study of it.

TABLE 1.1. Numbers of species of lizards in different families found in sympatry on desert study sites on three continents.

| LIZARD FAMILY | NORTH AMERICA | KALAHARI | AUSTRALIA |
|---|---|---|---|
| Agamidae | | 1 (1) | 2–8 (11) |
| Chameleontidae | | 1 (1) | |
| Gekkonidae | 1 (1) | 4–7 (7) | 5–9 (13) |
| Helodermatidae | 1 (1) | | |
| Iguanidae | 3–8 (9) | | |
| Lacertidae | | 3–5 (7) | |
| Pygopodidae | | | 1–2 (3) |
| Scincidae | | 3–5 (6) | 6–18 (28) |
| Teiidae | 1 (1) | | |
| Varanidae | | | 1–5 (5) |
| Xantusidae | 1 (1) | | |
| Total | 4–11 (13) | 12–18 (22) | 18–42 (61) |

NOTE: Total number of different species in each family in parentheses.

Turnover in species composition between study areas, as estimated by coefficients of community similarity among all possible pairs of communities within each continental desert system (Pianka 1973), is distinctly lower within western North America and in the Kalahari semidesert than it is in the Great Victoria desert. On the average, Kalahari study areas are closer together than are study sites in North America and Australia. In the North American deserts, community similarity values are high and rather uniform (mean = 0.67, standard error = 0.019, $N$ = 66) as they are in the Kalahari (mean = 0.67, S.E. = 0.015, $N$ = 66), indicating relatively little difference between study areas in species composition. In contrast, community similarity values are lower and more variable in the Great Victoria desert (mean = 0.49, S.E. = 0.027, $N$ = 28). Thus the "between-habitat" (MacArthur 1965) horizontal component of diversity is greater in Australia than in the other two continental desert-lizard systems (i.e., turnover in species composition between study sites is greater).

In all three continental desert-lizard systems, the same fifteen basic microhabitats were recognized: subterranean (for unknown, presumably largely historical, reasons, there are no subterranean lizards in the North American deserts), open sun, open shade, grass sun, grass shade, bush sun, bush shade, tree sun, tree shade, other sun, other

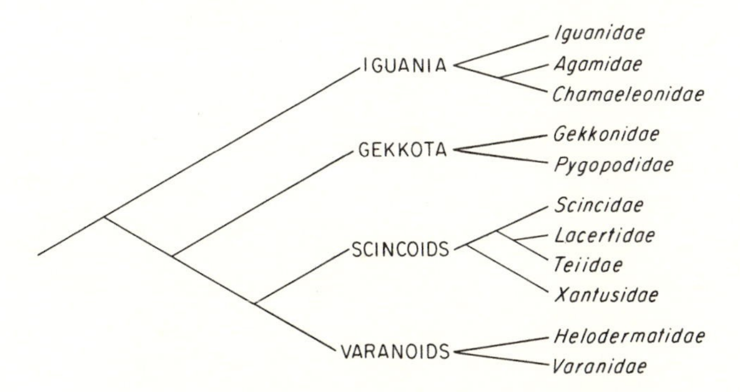


FIGURE 1.4. A crude cladogram of probable phylogenetic relationships among the eleven families of lizards found on study sites considered here, based on information presented by Estes (1983).

shade, low sun (within 30 cm. above ground), low shade, and high sun and high shade (over 30 cm. above ground). For some purposes, finer microhabitat resource states were used. Lizards at an interface between two or more microhabitats were usually assigned fractional representation in each. Only undisturbed lizards were used in these analyses. Considerable fidelity in microhabitat utilization is evident. Many species separate out using just these fifteen very crude microhabitat categories: for example, some species frequent open spaces between plants to the virtual exclusion of other microhabitats, whereas other species stay much closer to cover. These data are considered in greater detail in Chapters 3 and 8.

Because time of activity shifts seasonally with changes in ambient temperature (see also Chapter 3), it is useful to express times of activity either as time since sunrise or time since sunset (using sun time also corrects for longitudinal effects and time zones). Although time is a continuous variable, it is often more convenient to use discrete time intervals. For diurnal species, I used fourteen hourly categories (not all were sampled equally intensively), but limitations on human endurance dictated a maximum of only eight hourly categories for nocturnal species (again not sampled equally intensively). Any bias introduced through unequal sampling and/or the shorter night-time sampling period should be roughly similar on different areas and among the three continents.

In analysis of stomach contents, the same twenty very crude prey categories were distinguished on all continents (letters code prey categories shown in Figures 8.2, 8.3, and 8.4, and in Appendix E): centipedes (Ce); spiders (Sp); scorpions (Sc); solpugids (So) (absent from Australia); ants (A); wasps and other non-ant hymenopterans (W); grasshoppers plus crickets (G); roaches (Blattids) (B); mantids and phasmids (M); adult Neuroptera (ant lions) (N); beetles (Coleoptera) (Co); termites (Isoptera) (I); bugs (Hemiptera plus Homoptera) (H); flies (Diptera) (D); butterflies and moths (Lepidoptera) (Lp); insect eggs and pupae (E); all insect larvae (Lv); miscellaneous arthropods not in the above categories including unidentified ones (U); all vertebrate material including carrion and sloughed lizard skins (V); and plant materials (floral plus vegetative) (P). Both volumes and numbers of prey items were estimated for each of these categories (the resulting data consist of nearly half a million individual prey items).

As is true of microhabitat utilization, considerable consistency in diet is evident among species. For example, some species eat virtually nothing but termites, whereas others never touch them. Moreover, diets of many species change little in space or time (see also Chapter 10). Indeed, using just these twenty very crude prey categories allows reasonably clean separation of many pairs of lizard species on the basis of foods eaten (see also Chapters 4 and 8). When prey items are analyzed by either number or size, separation is substantially less than when the proportional representation of each food category by volume is used. For some purposes, much finer prey categories could be used. Termites were identified to species and caste for Kalahari lizards. Similarly, prey were identified to the finest possible categories for the 1978–79 Australian data set (for example, ants and termites were placed into size and/or color categories by family to generate some 97 ant and 58 different termite resource states). (Note that these very detailed data are only of limited utility for comparative purposes.)

A major virtue of these data is that identical methods and resource categories were used by the same investigator for each of three continental desert-lizard systems, enabling meaningful intercontinental comparisons.[4] This unique body of data thus allows fairly detailed analyses of patterns of resource utilization and community structure in these historically independent saurofaunas. Moreover, both dietary and microhabitat niche breadths and overlaps can be estimated; and species diversities and the spectra of resources actually exploited by entire lizard faunas varying widely in number of species can be ex-

[4] My research was *not* connected with, nor was it funded in any way by, the International Biological Program (I.B.P.). Indeed, any funding I managed to acquire was obtained in head-on competition with this program!

TABLE 1.2. Numbers of species of lizards with various basic modes of life.

| | NORTH AMERICA | | | KALAHARI | | | AUSTRALIA | | |
|---|---|---|---|---|---|---|---|---|---|
| MODE OF LIFE | *Mean* | *Range* | % | *Mean* | *Range* | % | *Mean* | *Range* | % |
| Diurnal | 6.3 | (4–9) | 86 | 8.2 | (7–10) | 56 | 17.0 | (9–25) | 60 |
| Ground-dwelling | 5.4 | (4–7) | 74 | 6.3 | (5.5–7.5) | 43 | 15.4 | (9–23.5) | 54 |
| Sit-and-wait | 4.4 | (3–6) | 60 | 2.4 | (1.5–2.5) | 16 | 5.3 | (2–7) | 18 |
| Widely foraging | 1.0 | (1) | 14 | 4.0 | (3–6) | 27 | 10.1 | (4–12) | 36 |
| Arboreal | 0.9 | (0–3) | 12 | 1.9 | (1.5–2.5) | 13 | 2.7 | (0–5.5) | 9 |
| Nocturnal | 1.0 | (0–2) | 14 | 5.1 | (4–6) | 35 | 10.2 | (8–13) | 36 |
| Ground-dwelling | 1.0 | (0–2) | 14 | 3.5 | (3–5) | 24 | 7.6 | (6–9) | 27 |
| Arboreal | 0 | — | 0 | 1.6 | (0.5–2.5) | 11 | 2.7 | (1–4) | 9 |
| Subterranean | 0 | — | 0 | 1.4 | (1–2) | 10 | 1.2 | (1–2) | 4 |
| All ground-dwelling | 6.4 | (4–8) | 88 | 9.8 | (9–11) | 67 | 23.0 | (15–31.5) | 78 |
| All arboreal | 0.9 | (0–3) | 12 | 3.5 | (2–5) | 24 | 5.4 | (1–9) | 18 |
| Total | 7.4 | (4–11) | 100 | 14.7 | (11–18) | 101 | 29.8 | (18–42) | 100 |

NOTE: Semiarboreal species are assigned half to arboreal and half to terrestrial categories.

amined. Still another dimension that can often be profitably exploited is area-to-area variation in species utilization patterns.

The ecological composition of the saurofaunas of these sites is summarized in Table 1.2. Numbers of species with different modes of life differ among sites within continents as well as between continental systems. More than twice as many species of diurnal ground-dwelling lizards occur on a typical Australian site than on an average site in the North American and Kalahari deserts; however, when expressed as a percentage of the total lizard fauna, diurnal ground-dwelling species constitute a full 74% of the North American saurofauna, compared with only 43% of the Kalahari lizard fauna and 51% of the lizard species on a typical Great Victoria desert site. Intercontinental variation in the absolute number of species that forage by sitting and waiting for their prey (see also Chapter 4) is slight, but the absolute number of widely foraging species increases rather markedly from North America to the Kalahari to Australia (see Table 1.2). When expressed as a percentage of the total saurofauna, only about 16% to 18% are diurnal sit-and-wait foragers in the southern hemisphere deserts (an average of 2.4 species in the Kalahari and 5.3 species in Australia), whereas a full 60% of the North American lizards (average 4.4 species) fall into this catgegory due to the low diversity of the latter desert-lizard system. The percentage contribution of all diurnal species, both ground-dwelling and arboreal, to the total fauna declines as the number of lizard species increases (i.e., nocturnality increases).

Arboreal, subterranean, and nocturnal species are conspicuously more prevalent in the two southern hemisphere deserts than they are in North America; subterranean lizards and arboreal nocturnal species are entirely missing from the North American saurofaunas, yet contribute three or four species to an average site in the two southern hemisphere deserts (Table 1.2). Numbers of species of arboreal lizards and their percentage contribution to total lizard faunas both tend to increase with number of lizard species (Table 1.2). However, arboreal lizard species are less well represented on structurally simple sites with low lizard diversity, even within the very diverse Australian saurofaunas.

The heightened relative importance of nocturnality among lizards in the Kalahari and Great Victoria deserts could be a consequence of any or all of at least three different historical factors: (1) Various desert systems could differ in diversity and abundance of available resources at night, such as nocturnal insects. (2) In North America, other taxa such as spiders might fill the ecological role of arboreal nocturnal lizards. Differences in species numbers and/or densities of insectivorous and carnivorous snakes, birds, and mammals might also play a crucial role. (3) Effects of the Pleistocene glaciations are generally acknowledged to have been stronger in the northern hemisphere, which must certainly have influenced the evolution of nocturnal lizards. However, in at least the southernmost parts of the Sonoran desert of North America, present-day climates (see Chapter 2) are easily adequate for nocturnal lizards. The eublepharine gecko *Coleonyx* thrives as a nocturnal terrestrial lizard in the Sonoran and Mojave deserts, occurring as far north as latitude 37 degrees (some geckos reach comparable southern latitudes in Australia). As expressed so forcefully by Hutchinson (1959) in a slightly different context, "if one . . . species can [exist in the nocturnal terrestrial niche], why can't more?" The absence of an arboreal gecko from the flatland deserts in the southern Sonoran desert is also most puzzling, particularly since the rock-dwelling climbing gecko *Phyllodactylus xanti* occurs nearby. A tree or shrub-climbing gecko species might well be able to invade this desert region if given an opportunity. Without such an introduction, either accidental or deliberate, all such interpretations must unfortunately remain speculative.

Various historical factors, such as degree of isolation and available biotic stocks (particularly those of potential prey, predators, and competitors) have clearly shaped these lizard faunas in other ways. Certain ecological roles occupied by nonlizard taxa in North America and the Kalahari have been usurped by Australian desert lizards. Thus Australian pygopodid and varanid lizards clearly replace certain snakes and mammalian carnivores, taxa that are impoverished in Australia. Numbers of species in various vertebrate taxa found on study sites are summarized in Table 1.3. There are more species of snakes on sites in the North American deserts than there are on the other continents, although species richness of Australian "snakes" becomes similar when adjusted by addition of legless snake-like pygopodid lizards to the real snakes. Just as pygopodids appear to replace snakes, varanid lizards in Australia are clearly ecological equivalents of carnivorous mammals such as the kit fox and coyote in North America. Mammal-like and snake-like lizards contribute from one to eight (usually only four) species on various Australian study sites and thus represent a relatively minor component of the overall increase in numbers of lizard species on that continent, however. Nevertheless, such usurpation of the ecological roles of other taxa has clearly expanded the variety of resources (or "overall niche space") exploited by Australian desert lizards.

In addition to such conspicuous replacements of one taxon by another, more subtle competitive interactions between taxa doubtlessly occur, particularly between lizards and insectivorous birds. Fewer species of ground-dwelling insectivorous birds occur in Australia than in the Kalahari (Table 1.3), which may reduce competition between lizards and birds in Australia (Pianka and Huey 1971; Lein 1972). With increases in the total number of species of birds plus lizards, the number of lizard species increases faster than bird species in Australia, whereas in North America and the Kalahari, bird species density increases faster than lizard species density (Figure 1.5; Pianka 1971e, 1973). Reasons

TABLE 1.3. Species densities of various vertebrates on desert study areas.

| | NORTH AMERICA | | KALAHARI | | AUSTRALIA | |
|---|---|---|---|---|---|---|
| TAXON | *Mean* | *Range* | *Mean* | *Range* | *Mean* | *Range* |
| All lizards | 7.4 | (4–11) | 14.7 | (11–18) | 29.8 | (18–42) |
| Pygopodid lizards | — | — | — | — | 1.7 | (0–3) |
| All birds | 7.8 | (3–16) | 22.8 | (15–40) | 28.3 | (15–41) |
| Ground-foraging insectivorous birds | — | — | 7.3 | (4.5–11.7) | 4.8 | (2.8–5.8) |
| Snakes | 4.5 | (2–9) | 2.2 | (1–6) | 3.6 | (1–6) |
| Small mammals | 5.3 | (4–8) | — | — | 1.5 | (1–3) |
| Site totals | 24.7 | (14–40) | 40.2 | (27–63) | 64.8 | (36–88) |

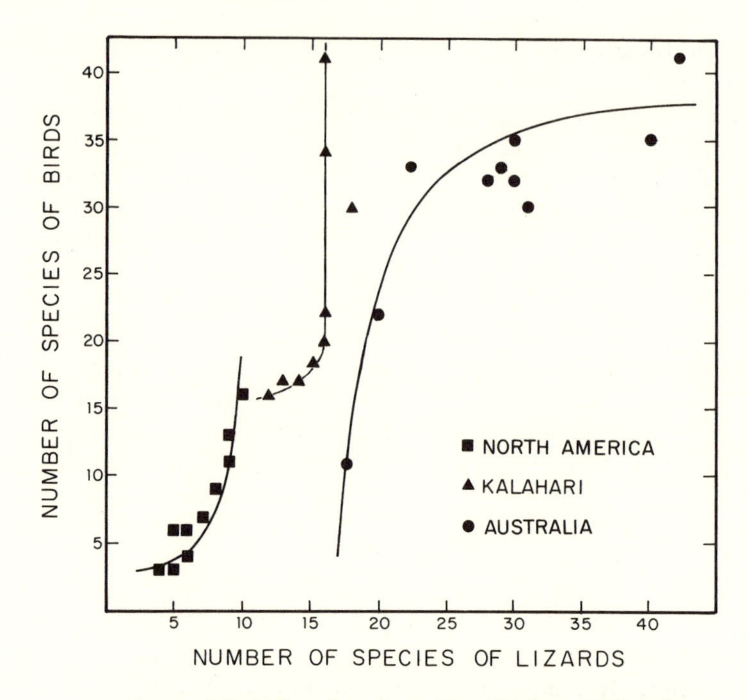


FIGURE 1.5. A plot of total number of bird species found on any given study site versus the total number of species of lizards recorded at the same site. Note that lizard species are added faster than bird species in Australia, whereas the reverse is true in North America and the Kalahari.

for these differences among continents are elusive, but one factor may be that very few birds migrate in Australia, both because of its isolation and the limited areas at high latitude. In contrast, a fair number of migrant bird species periodically exploit the North American and African deserts; lizard faunas of the latter two desert systems could well be adversely influenced by intermittent but still regular competitive pressures from these insectivorous avian migrants.

The impact of prey taxa on the composition of saurofaunas is illustrated by the conspicuous and diverse termite fauna of southern Africa, which has apparently facilitated the evolution of termite-specialized species of skinks, lacertids and geckos (Huey et al. 1974; Huey and Pianka 1974; Pianka et al. 1979; Pianka and Huey 1978). Indeed, termites comprise a full 41% by volume of the diet of Kalahari lizards, but constitute only about 16% of the overall diet of the entire saurofauna in western North America and only about 18.5% of the diet of Western Australian lizards (see also Chapters 4 and 8).

Numerous more elusive interactions between taxa doubtlessly occur. For instance, one reason that the Australian deserts support such rich lizard faunas could involve reduced predation pressures from snakes, raptors, and/or carnivorous mammals on that isolated continent (however, many Australian lizards, both varanids and pygopodids, do prey upon other lizards). Similarly, the higher incidence of arboreal and nocturnal lizard species in the Kalahari and Australia compared to North America could well be related to fundamental differences in the niches occupied by other members of these communities, including potential competitors among arthropods, snakes, birds, and/or mammals.

The possibility that soil fertility might differ fundamentally between Australia and southern Africa also exists (Charley 1972; Charley and Cowling 1968; Milewski 1981). Milewski has suggested that, if so, the putatively poorer Australian soils might, by virtue of their low productivity, favor ectotherms and thus tend to support a more diverse lizard fauna. Summer rainfall regimes and long stabilized red sandridges characterize both the Kalahari desert and the Great Victoria desert. Indeed, in their basic physical aspects (topography, geomorphology, and climatology), the two regions are so similar that, without recourse to their different floras and faunas, one would be hard pressed to distinguish between the two continents. Major soil fertility differences would therefore seem rather unlikely. Indeed, the only published comparison of nutrient concentrations and pH of selected soil samples from the southern Kalahari versus those in central Australia was made by Buckley (1981b), who reports somewhat lower percentages of organic material in Australia, but higher concentrations of calcium, potassium, and phosphorus. As pointed out by Buckley, nutrients vary considerably from interdunal streets to sandridge crests, making such crude comparisons between continents of dubious utility. Nonetheless, closer scrutiny of Milewski's somewhat simplistic hypothesis, accompanied with hard data and more detailed intercontinental comparisons, would certainly be of interest.

The physical structure of the vegetation profoundly influences the composition of lizard faunas as well. The mere existence of the unique hummock-grass plant life form in Australia (Beard 1976; Burbidge 1953) is a major factor contributing to lizard diversity on that continent (Pianka 1969a, 1969b, 1973, 1975, 1981). Area N, a pure "spinifex" (*Triodia*) grass flat supports at least 16 species of lizards (perhaps as many as 20), including 6 or 7 species of *Ctenotus* skinks. These grass tussocks are extraordinarily well suited for lizard inhabitants, providing not only protection from predators and the elements, but a rich insect food supply as well. Certain lizard species appear to spend almost their entire lives within dense *Triodia* clumps, while other species exploit their edges. Still other lizards forage in the open spaces between tussocks but rely on spinifex clumps for escape cover in emergencies. The latter species tend to have relatively longer hind legs than the former (see Chapter 11) and are presumably faster runners; a trade-off exists, however, since long-legged open-dwelling species must move clumsily through dense vegetation whereas short-legged species literally "swim" through it with ease.

Lizard faunas of shrub desert sites on each of the three continents are compared in Table 1.4. The structure of the vegetation, consisting of low microphyllous chenopod shrubs, is virtually identical on these three sites, which were chosen to control vegetative structural complexity (plants of the globally distributed genus *Atriplex* occur on all three areas). On North American sites, only 5 species of lizards are present (a sixth species is added in the south). The Kalahari site supports 13 lizard species whereas an Australian shrubby area in the dry lakebed of Lake Yeo presumably sustains a full 18 species (Pianka 1969a, 1971e, 1981). The major differences between continents are traceable to nonlizard-like lizards and to nocturnal species. Four gecko species are nocturnal in the Kalahari while eight species of geckos and skinks are active at night on the Australian site (only one nocturnal lizard species exists on southern North American sites). Insect-like species (*Mabuya variegata* and *Menetia greyi*), a mammal-like lizard (*Varanus gouldi*), and a "worm-like" subterranean skink (*Lerista muelleri*) further expand the lists in the southern hemisphere. Numbers of species of truly lizard-like lizards that are both diurnal and terrestrial (or semiarboreal) are much more comparable among the three continents: North America (5 species), Kalahari (8 species), and Australia (7 species).

Effects of historical variables such as the Pleistocene glaciations could also be considerable, but are exceedingly difficult to evaluate. The North American deserts are generally acknowledged to be of relatively recent geologic origin (Axelrod 1950), although subdesert conditions must have prevailed in the general region long before the origin of true deserts. During the upper Tertiary, American deserts expanded but then became restricted to northern Mexico and the extreme southwestern United States with the onset of the Pleistocene glaciations. Presumably these deserts expanded rapidly to their present boundaries with the retreat of the glaciers about 10,000 years ago. The sands of the Kalahari are largely of aeolian origin and were formed and originally distributed during the Tertiary; but later they were

TABLE 1.4. Lizard faunas on three chenopodeaceous shrubby sites with very similar vegetative structures.

| NORTH AMERICA | KALAHARI | AUSTRALIA |
|---|---|---|
| *Cnemidophorus tigris* | *Mabuya occidentalis*<br>*Eremias lugubris*<br>*Eremias namaquensis* | *Ctenotus schomburgkii*<br>*Ctenotus leonhardii* |
| *Uta stansburiana* | *Eremias lineo-ocellata* | *Ctenophorus isolepis* |
| *Phrynosoma platyrhinos* | *Agama hispida (?)* | *Moloch horridus* * |
| *Crotaphytus wislizeni* | *Mabuya striata* | *Ctenophorus inermis*<br>*Ctenophorus reticulatus* |
| *Callisaurus draconoides* | *Meroles suborbitalis*<br>*Ichnotropis squamulosa* | *Ctenophorus scutulatus* |
| | *Mabuya variegata* | *Menetia greyi* *<br>*Lerista muelleri* * |
| | | *Varanus gouldi* |
| *Coleonyx variegata* | *Colopus wahlbergi* | *Rhynchoedura ornata* |
| | *Ptenopus garrulus* | *Diplodactylus conspicillatus* |
| | *Chondrodactylus angulifer* | *Nephrurus vertebralis* |
| | *Pachydactylus capensis* | *Heteronotia binoei* |
| | | *Diplodactylus strophurus* |
| | | *Gehyra variegata* |
| | | *Egernia inornata* |
| | | *Eremiascincus richardsoni* |

NOTES: The globally distributed plant genus *Atriplex* occurs on all three sites. Ecological equivalents that are crudely approximate are aligned horizontally. Nocturnal species are listed in the bottom half of table. *Agama* is a very questionable counterpart of *Phrynosoma* and *Moloch*.
* Not actually collected on the area, but highly expected to occur there based on autecological considerations and occurrences on other areas.

redistributed in the Pleistocene by both wind and water. Lancaster (1979) presents evidence for a widespread humid period in the Kalahari during the late Pleistocene. Kalahari sandridges probably assumed roughly their present distribution during drier periods over the last 10,000 years; the subsequent stabilization of the sandridges by vegetation suggests that a slight amelioration of the climate may have taken place in more recent times. So-called "Kalahari" sands are widespread in southern Africa, occurring well beyond the confines of the currently recognized Kalahari semidesert. Dry to very dry conditions have probably prevailed in most of this area since the middle to the end of the Tertiary (Lancaster 1984), although some workers have suggested otherwise. Views as to the age and history of the Australian deserts are also varied and conflicting. Crocker and Wood (1947) postulated a "Great Aridity" during the Pleistocene following the last glaciation, but others have argued for a much greater antiquity for at least some portions of the continent (Bowler 1976; Bowler et al. 1976; Galloway and Kemp 1981). Certainly Australian sandridges appear to be exceedingly ancient (Bowler 1976).

Appropriately selected "natural experiments" may actually allow a limited measure of "control" over historical phenomena such as the glaciations. Differences between independently evolved faunas occurring in areas with comparable present-day climates (see next chapter) and vegetation structures presumably reflect such historical factors (Orians and Solbrig 1977). Natural variation is also frequently very useful because experimental manipulation of ecological systems is often extremely difficult and therefore impractical, if not immoral, illegal, or even actually impossible. For example, a fruitful natural experiment on competition may exist when a species occurs both with and without a potential competitor in similar habitats in different parts of its geographic range. Such situations have been considered analogous to removal-addition experiments, and, if chosen and studied with enough care, niche shifts can sometimes be related to the underlying effects of interspecific competition. Lizards have contributed greatly to such comparative descriptive studies (Huey et al. 1974; Huey and Pianka 1974, 1977b; Schoener 1975; Schoener et al. 1979), although interpretation of data is inevitably fraught with difficulties (Colwell and Fuentes 1975; Connell 1975; Grant 1972, 1975, and above papers). Similarly, as indicated above, comparing historically independent but otherwise basically comparable ecological systems can sometimes help ecologists to attempt to assess the predictability of evolutionary pathways (Recher 1969; DiCastri and Mooney 1973; Orians and Solbrig 1977; Schall and Pianka 1978).

Organisms that fill similar ecological niches in different, independently evolved biotas are termed "ecological equivalents" (Grinnell 1924). Some such convergent evolutionary responses of lizards to the desert environment, although imperfect, are evident between the three continents (Pianka 1975, 1985). For example, the Australian and North American deserts both support a cryptically colored, thornily armored, ant specialized species: the agamid *Moloch horridus* exploits this ecological role in Australia (Pianka and Pianka 1970), while the iguanid *Phrynosoma platyrhinos* occupies it in North America (Pianka and Parker 1975). No Kalahari lizard has adopted such a life style. Both North America and Australia also have long-legged species that frequent the open spaces between plants (*Ctenophorus scutulatus* and *C. isolepis* in Australia: Pianka 1971c, d; *Callisaurus draconoides* in North America: Pianka and Parker 1972), as well as a medium-sized lizard-eating lizard (*Varanus eremius* in Australia: Pianka 1968; and *Crotaphytus wislizeni* in North America: Parker and Pianka 1976). A few Kalahari-Australia species pairs are also crudely convergent: for example, the subterranean skinks *Typhlosaurus* and *Lerista* are roughly similar in their anatomy and ecology, as are the semiarboreal agamid lizards *Agama hispida* and *Pogona minor*. Although I once suggested that several pairs of species of nocturnal geckos were vaguely similar primarily on the basis of superficial morphological similarities (Pianka 1975), a more objective analysis showed little ecological similarity among those four pairs (Pianka and Huey 1978); we did however, identify several other pairs of approximate ecological equivalents based on similar patterns of resource utilization. Relatively few convergences are apparent among all three continents (Pianka 1975). Ecologies of putatively convergent species pairs inevitably differ markedly when scrutinized closely (Pianka and Pianka 1970; Pianka 1971c). In fact, the *differences* in the ecologies of most lizard species among these three continental desert-lizard systems are much more striking than are the similarities. It is easy to make too much out of convergence, and one must always be wary of imposing it upon the system under consideration.

# 2 Productivity, Climatic Stability, and Predictability

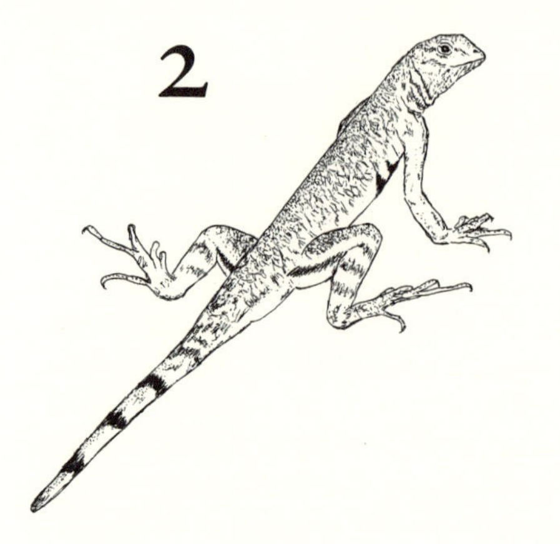

In deserts, sunlight is almost always present in excess, but water is usually in short supply. As a result, water is recognized as a "master limiting factor" and photosynthesis is sharply limited by water availability. On fairly flat desert areas with little topography and moderate runoff (such as my study sites), net annual primary production is directly proportional to annual precipitation (Blaisdell 1958; Pearson 1965; Rosenzweig 1968; Walter 1939, 1955, 1964). This strong positive correlation between precipitation and productivity is extremely convenient, for it allows primary production to be estimated from rainfall records of weather stations near study sites. Temporal variation in primary productivity should likewise be reflected in patterns of precipitation through time. Precipitation data can thus be exploited to generate estimates of average annual productivity as well as the variability in productivity both seasonally and on a long-term basis over a period of years. Similarly, recent precipitation statistics can be used as short-term indicators of conditions in the immediate past.

Provided that insect populations in turn reflect conditions for plant growth, climatic statistics become useful crude estimators of both the absolute amount and the variability of food supplies for desert lizards. In various studies, recent precipitation has been correlated with insect abundances, diversity of foods eaten by lizards, lizard clutch sizes, clutch frequency, growth rates and body size, fat bodies, and lizard population densities (Case 1976, 1982; Dunham 1980; Hoddenbach and Turner 1968; Mayhew 1967; Pianka 1970c; Parker and Pianka 1975; Turner et al. 1969). Evidently, one may safely surmise that rainfall often determines food supplies, which in turn regulate growth and reproduction and, ultimately, lizard population densities.

Among my study sites, long-term average annual precipitation varies from a minimum of 9.3 cm. on the North American area P to maxima of 28.6 cm. and 31.2 cm. on the Kalahari area T and the Australian "Red Sands" area, respectively. Standard deviation in total annual precipitation varies from 4.4 to 17.7 cm. (Appendix B), and is positively correlated with the mean (see also below). Although scatter is extensive and neither correlation need necessarily have biological meaning, number of lizard species is significantly correlated with both long-term mean annual precipitation ($r = .61$, $P < .001$) as well as with the standard deviation in annual precipitation ($r = .63$, $P < .001$; Figure 2.1). In the Kalahari semidesert, however, the number of lizard species is weakly *negatively* correlated with long-term average annual precipitation ($r = -.45$, $P < .15$) and is significantly *negatively* correlated with standard deviation in precipitation ($r = -.88$, $P < .01$; see also Figure 2.1). When standard deviation in precipitation is expressed as a fraction of the annual mean (i.e., as the coefficient of variation), there is no correlation with number of lizard species, either within any continent or among all three.

Several other related measures of precipitation variability have been used by other workers. The "predictable amount of precipitation," the

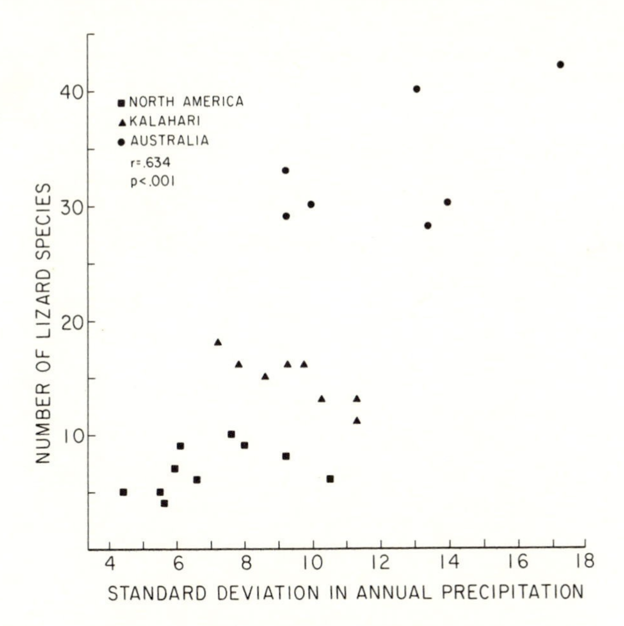


FIGURE 2.1. Plot of number of lizard species versus the standard deviation in annual precipitation. See text.

mean minus the standard deviation, is positively correlated with number of small mammal species in North American deserts (Brown 1975). For my sites, this measure does not correlate with lizard diversity but is strongly correlated with the simpler measure of mean annual precipitation ($r$ = .86). The ratio of the variance in annual precipitation over the mean squared is used by Rappoldt and Hogeweg (1980) as a measure of "environmental variability." For my sites, this measure is virtually identical to the coefficient of variation in annual precipitation ($r$ =.99). Neither of these metrics correlates with either the number of lizard species or with lizard species diversity.

As noted above, long-term mean and standard deviation in annual precipitation are themselves positively correlated ($r$ =.77, $P < .001$). If the effects of long-term average annual precipitation are held constant by partial correlation, lizard species numbers and standard deviation in precipitation remain significantly correlated ($r$ =.52, $P <$ .01). However, when standard deviation in precipitation is partialed out, number of lizard species no longer correlates with long-term average annual precipitation ($r$ =.11). Hence productivity alone may not promote diversity, but annual variability in productivity is perhaps more likely (partial correlation will not separate variables unequivocally). Certainly, productive sites might be expected to favor narrower diets (this can be tested directly—see pp. 53–54) and as a result might be expected to support more species of lizards than less productive areas, but the mechanisms by which increased variability in itself promotes diversity are perhaps less obvious. Temporal heterogeneity might actually facilitate coexistence, both by continually altering the relative competitive abilities among members of a community and by periodic reductions in population sizes that would lower the intensity of competition (Hutchinson 1961; Chesson 1982, 1983, 1985; Chesson and Case 1985; Chesson and Warner 1981). Such a disturbance mechanism probably operates in desert lizard faunas. However, as noted above, within the Kalahari the trend would seem to go in the opposite way (Figure 2.1).

Summer rains characterize the Sonoran desert in the North American system, the southern Kalahari, and the Western Australian desert (Figure 2.2). There is essentially no winter precipitation in the Kalahari, but some does occur in both the Great Victoria and the North American deserts; the vast majority of precipitation in the Mojave desert (in North America) falls during winter (Figure 2.2). Some of this precipitation is probably relatively unavailable to desert plants because of low potential evapotranspiration during winter. If so, rainfall statistics would overestimate productivity for the Mojave. In the Great

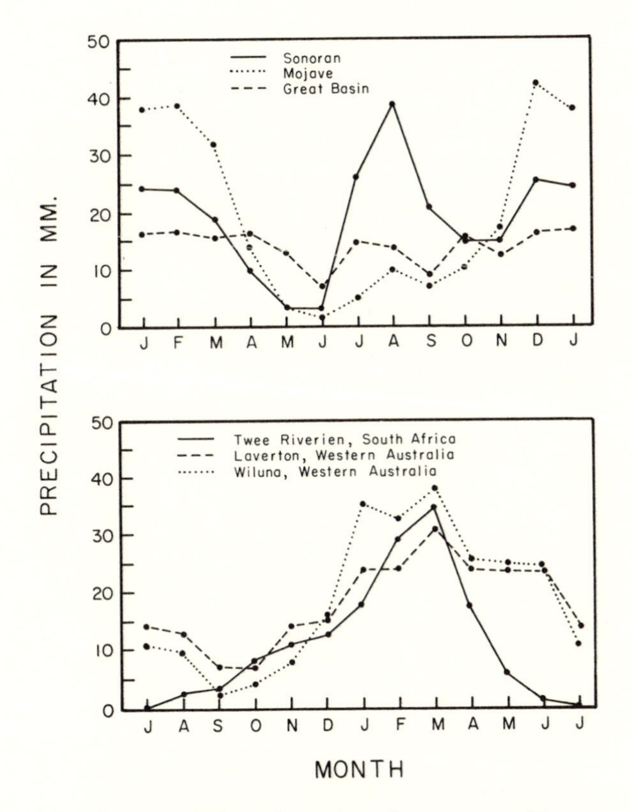


FIGURE 2.2. Annual marches of average monthly precipitation characteristic of various desert regions. Note the pronounced similarity between the Kalahari and Great Victoria deserts.

Basin desert, precipitation is spread fairly evenly over the year with little periodicity (Pianka 1967).

Seasonal predictability of precipitation (and productivity), as manifest by time-series analysis of monthly precipitation (Figure 2.3), is somewhat similar in warm deserts on the three continents, although there is some biologically interesting variation within the North American desert system (Pianka 1967). Autocorrelation coefficients of annual precipitation range from −.59 to +.32 among North American sites, from −.38 to +.14 in the Kalahari, and from −.23 to +.36 among six Australian sites with rainfall data (Figure 2.4). Number of lizard species tends to increase weakly as annual precipitation becomes more erratic and less predictable, both between and within continents (not, however, within Australia).

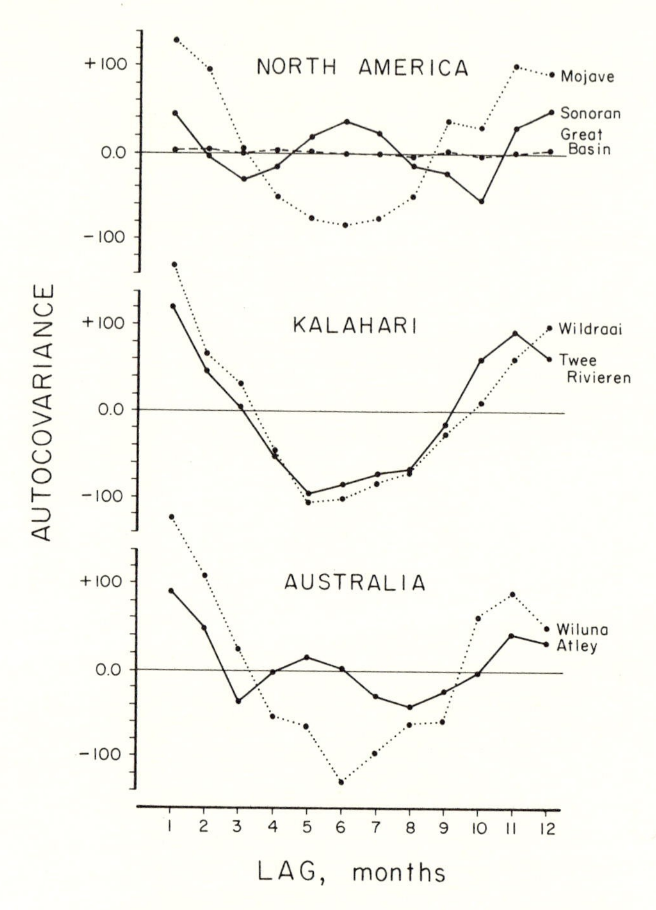


FIGURE 2.3. Time series analyses of monthly precipitation totals reveal periodicities. Autocovariances near zero reflect unpredictable conditions, whereas large positive autocovariances indicate that, at the lag time shown, conditions will tend to be similar to those at an earlier point in time. Negative autocovariances mean that conditions will usually differ.

A number of instructive population-level correlations with long-term precipitation statistics merit mention here (most are relevant to subsequent chapters). Among species I have studied, the most informative is the abundant North American whiptail, *Cnemidophorus tigris*. As in many lizards, this teiid species possesses fat storage organs that protrude into the body cavity from the ventral pelvic region. These "fat bodies" vary in size from area to area (Pianka 1970c), as well as

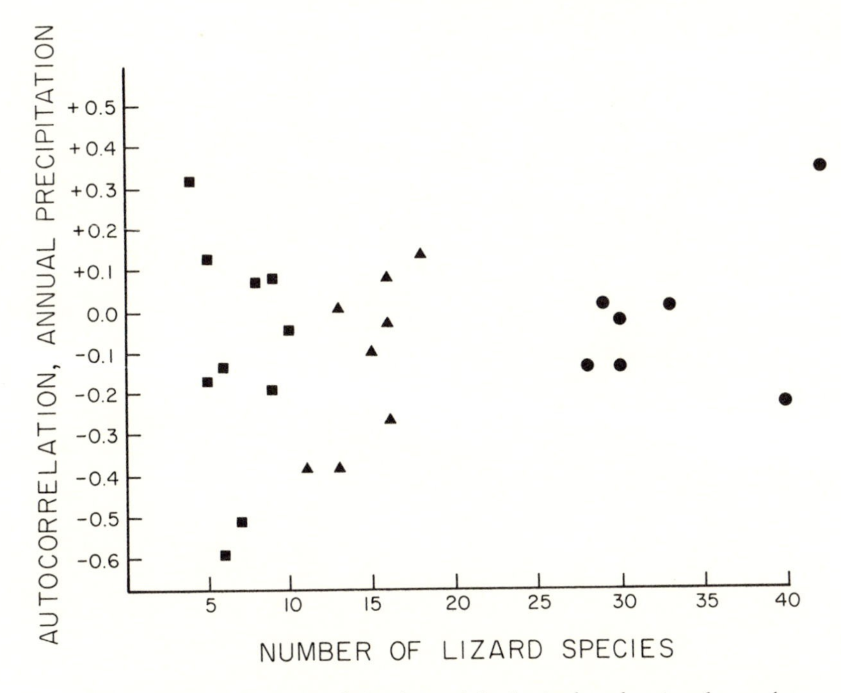


FIGURE 2.4. Annual autocorrelation in precipitation is plotted against the number of lizard species.

inversely with reproductive activities during the season. Fat bodies of whiptails from drier areas tend to be larger than those from wetter sites (Figure 2.5) rather than smaller, as might at first have been anticipated (the same phenomenon was reported for the chuckwalla *Sauromalus obesus* by Case 1976). A similar, though not so strong, inverse correlation exists with short-term measures of precipitation. Initially somewhat surprising, these provocative data suggest that whiptails from the less productive areas allow themselves a greater margin of safety, presumably due to the more probable occurrence of drought. This interpretation is also suggested by the greater range in mean fat body sizes on study areas with low long-term mean precipitation values. Such a "bet-hedging" mechanism would require a genetic basis to persist. Note that other explanations for a shift in energy budgets with changes in productivity are also possible. For example, increased predation on more productive areas could simply require greater energy expenditure on predator avoidance and thus result in reduced fat reserves.

Several other important features of the ecology of *Cnemidophorus*

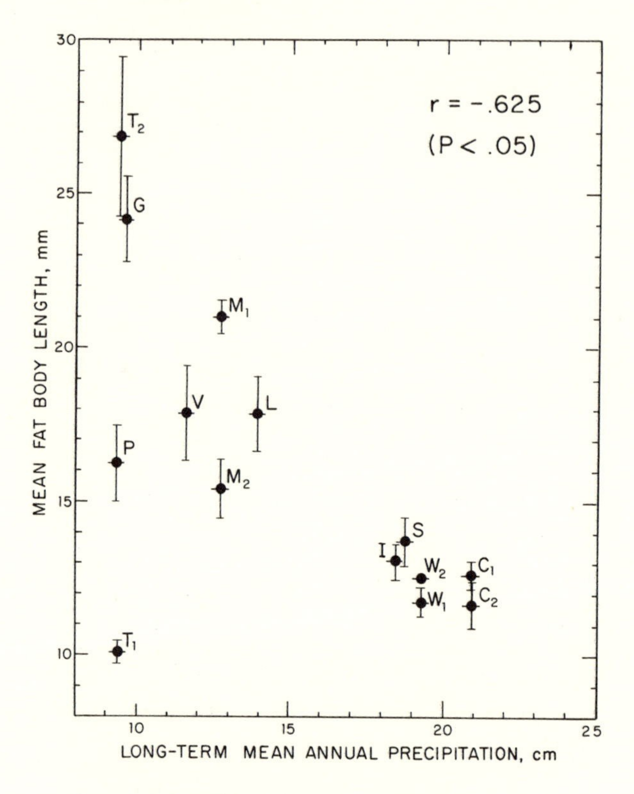


FIGURE 2.5. In the teiid lizard *Cnemidophorus tigris*, fat bodies tend to be larger (as well as more variable in size) on drier sites than they are on study areas with greater long-term average annual precipitation.

*tigris* are also sensitive to precipitation, including abundance, dietary diversity, and fecundity (Pianka 1970c). Abundance is positively correlated with both the long-term average annual precipitation ($r_s$ = .604, $P < .05$) and with the total precipitation during the five years prior to my estimate of abundance ($r_s = .561$, $P < .05$). The diversity of foods eaten by populations of *Cnemidophorus tigris* varies inversely with recent precipitation (Figure 4.4); optimal-foraging theory predicts narrower diets when food is abundant and broader diets as food becomes scarcer (see Chapter 4 and MacArthur and Pianka 1966). Clutch size in *Cnemidophorus tigris* varies substantially from year to year as well as from area to area. For example, mean clutch size on my study area T in 1963 (a dry year) was only 2.3, but on the same site in 1964 (an exceedingly wet year) the average clutch was 3.4 eggs, an increase of a full egg per female (this increase was *not* an artifact of female size since those collected in the second year were no larger than those collected in the first year). Whiptails from areas with less than the long-term average annual precipitation during the last five years tended to have smaller than average clutches, whereas those from areas with above-average rainfall had larger clutches (Figure 5.1).

The other primary component of climate is, of course, temperature. Like other ectotherms, lizards rely on basking, shade, and/or cool versus warm ambient air or substrates for heat gains and losses from their bodies: hence temperature relations are of utmost importance in understanding lizard activities. Deserts are well known for their thermal extremes. Just before dawn, temperatures are typically uniform and quite low but rise rapidly so that, by midafternoon, temperatures in many microhabitats regularly reach lethal limits. Lizards compensate somewhat for such profound temperature changes both by exploiting spatial heterogeneity in thermal conditions and by selecting temporal "windows" for activity that provide the most suitable ambient temperatures. Macroclimatic and seasonal thermal patterns are of primary concern here, but microclimatic and daily phenomena are examined further in the next chapter.

In the warm deserts, annual marches of average daily temperature are relatively similar in all three continental systems (Figure 2.6). The North American desert system has a more continental climate, however, and is characterized by a greater annual amplitude and much greater variation from place to place within the deserts. Because North America includes both "cold" and "warm" deserts, thermal variation within that continent is greater than it is between continental desert systems. Figure 2.7 shows the annual march of average daily range in temperature (mean maximum minus mean minimum daily temperature) for selected stations near study areas and typical of each of the three desert systems. Within the North American desert system (upper panel of Figure 2.7), daily range in temperature tends to increase with latitude, especially during the summer months. Average temperatures also tend to be lower in the north. An intercontinental comparison shows that, during winter but not during summer months, average daily range in temperature is somewhat greater in the Kalahari than in the Australian desert. In the southern Sonoran desert (upper panel of Figure 2.7) average daily range in temperature is low and quite comparable to Australia. Average monthly minimum temperatures, reflecting nighttime temperatures for nocturnal lizards, are quite similar in all three of the warm-desert systems (Kalahari, Great Victoria, and Sonoran deserts), and are, in fact, actually warmest in the Sonoran

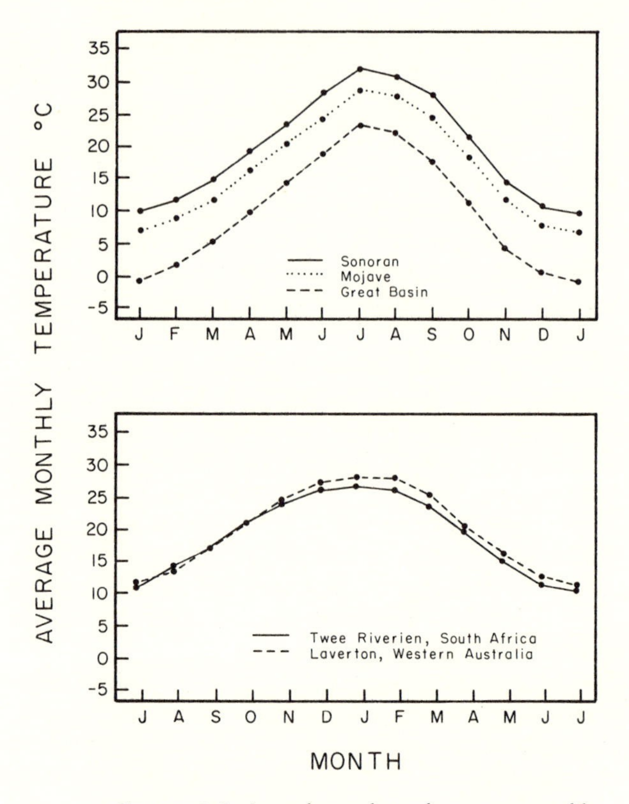


FIGURE 2.6. Annual marches of average monthly temperatures characteristic of each of the various desert regions.

desert (Figure 2.8).[1] Although they are cold, winter temperatures at midday are mild enough in the Kalahari and Australian deserts for some species of lizards to be active during all months of the year. In North America, however, only the southernmost Sonoran desert is warm enough for year-round lizard activity. Growing seasons in the Mojave and Great Basin deserts decrease with elevation and latitude; for example, in the far northern Great Basin desert at site I in southern Idaho, average length of the frost-free period is a mere 120 days.

[1] Recall that the North American deserts support a comparatively impoverished nocturnal lizard fauna (see p. 16).

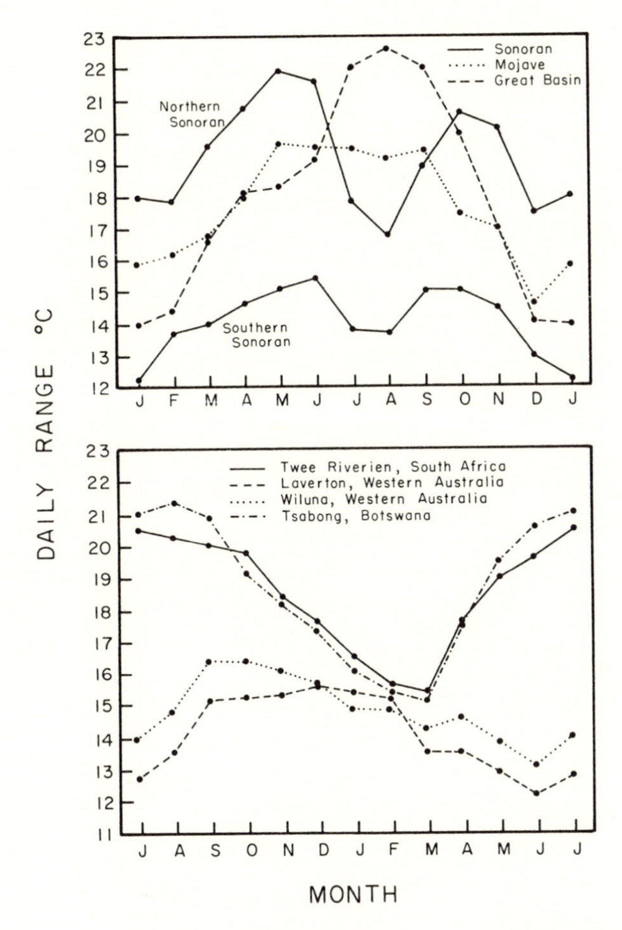


FIGURE 2.7. Annual marches of the average daily range in temperature for each of the various desert regions.

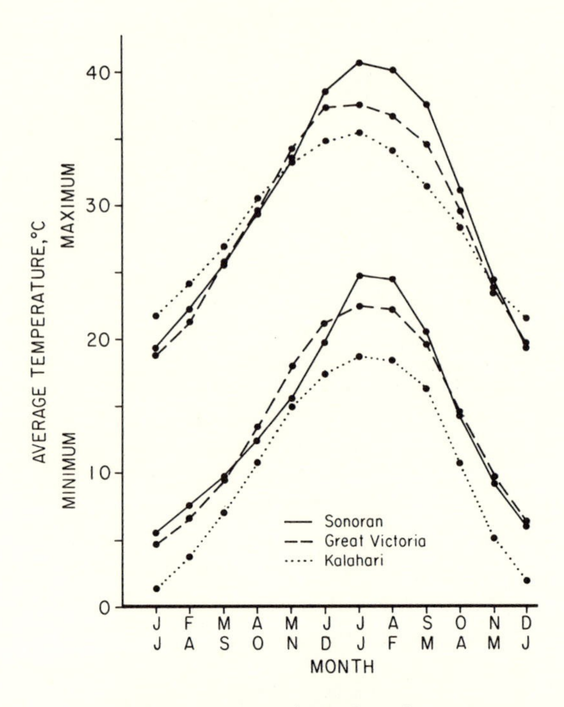


FIGURE 2.8. Annual marches of average maximum and average minimum temperatures for each of the various desert regions.

# 3 Thermal Relations, Spatial and Temporal Patterns of Activity

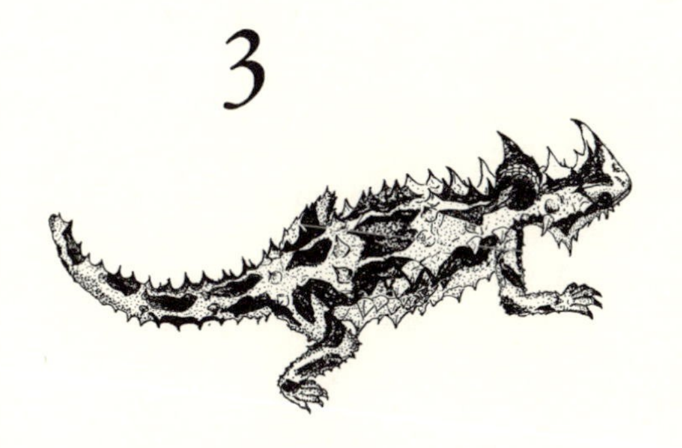

Even a casual observer quickly notices that various species of desert lizards differ markedly in their times and places of activity. Some are active early in the morning, but other species do not emerge until late morning or midday. Most geckos and pygopodids and some Australian skinks are nocturnal. Certain species are climbers, others subterranean, while still others are strictly surface dwellers. Among the latter, some tend to be found in open areas whereas others frequent the edges of vegetation. Because such spatial and temporal differences limit the frequency of encounters between species as well as expose them to differing food resources, any potential effects of interspecific competition would tend to be ameliorated. Indeed, avoidance of competition is perhaps the most plausible basis for the evolution and maintenance, and thus the very existence, of these differences. (Other possibilities, such as physiological design constraints and predator avoidance tactics, need also be considered [Toft 1985].)

In the North American deserts, zebra-tailed lizards (*Callisaurus draconoides*) are usually in the open sun when first sighted; in contrast, side-blotched lizards (*Uta stansburiana*) are most frequently found underneath shrubs. Other iguanid species, such as *Sceloporus magister* and *Urosaurus graciosus*, are almost always found in trees at some distance above ground. *Urosaurus* frequent smaller branches in the tree canopy where they capture most of their insect prey, whereas *Sceloporus* use tree trunks as perches from which they make forays to the ground to feed (such species perhaps should not be labeled "arboreal" or "semiarboreal," but rather should be called climbing ground-feeders). In any case, these lizards exploit a different microhabitat than true ground-dwelling species that forage at considerably greater distances away from trees; hence any potential for competition for food should be reduced by this differential use of space.

Microhabitat utilization frequencies for various species of lizards in each of the three continental desert systems are summarized in Appendix C. Table 3.1 lists overall frequencies of use of various microhabitat elements, along with total numbers of undisturbed lizards observed in each. Considerable intercontinental variation in the incidence of use of different microhabitats is evident: for example, in the two southern hemisphere deserts, substantially more lizards are arboreal and subterranean.

The diversity of microhabits used[1] by various species of North American desert lizards (also termed "microhabitat niche breadths"; see also Chapter 7) varies from 1.0 in the specialized night lizard, *Xantusia vigilis* (found only in the fallen rubble underneath Joshua trees), to 3.87 in the much more generalized side-blotched lizard, *Uta stans-*

TABLE 3.1. Microhabitats used by all lizards in each continental desert-lizard system.

| MICROHABITAT CATEGORY | NORTH AMERICA | | KALAHARI | | AUSTRALIA | |
|---|---|---|---|---|---|---|
| | *Number* | % | *Number* | % | *Number* | % |
| Subterranean | 0 | 0.0 | 579 | 12.1 | 50 | 0.8 |
| Terrestrial | 2826 | 96.0 | 3506 | 73.1 | 4788 | 78.1 |
| Open sun | 1335 | 45.3 | 890 | 18.6 | 1003 | 16.4 |
| Grass sun | 92 | 3.1 | 155 | 3.2 | 968 | 15.8 |
| Bush sun | 883 | 30.0 | 547 | 11.4 | 231 | 3.8 |
| Tree sun | 103 | 3.5 | 126 | 2.6 | 41 | 0.7 |
| Other sun | 95 | 3.2 | 6 | 0.1 | 29 | 0.5 |
| Open shade | 49 | 1.7 | 546 | 11.4 | 832 | 13.6 |
| Grass shade | 2 | 0.1 | 274 | 5.7 | 1126 | 18.4 |
| Bush shade | 165 | 5.6 | 765 | 15.9 | 334 | 5.5 |
| Tree shade | 30 | 1.0 | 179 | 3.7 | 141 | 2.3 |
| Other shade | 72 | 2.4 | 18 | 0.4 | 83 | 1.4 |
| Arboreal | 119 | 4.2 | 710 | 14.8 | 1291 | 21.1 |
| Low sun | 12 | 0.4 | 125 | 2.6 | 90 | 1.5 |
| Low shade | 6 | 0.2 | 109 | 2.3 | 487 | 7.9 |
| High sun | 50 | 1.8 | 200 | 4.2 | 127 | 2.1 |
| High shade | 51 | 1.8 | 276 | 5.8 | 587 | 9.6 |
| Total | 2945 | | 4795 | | 6129 | |

[1] Measurements based on proportional utilization using Simpson's (1949) index of diversity as an index of microhabitat niche breadth.

*buriana* (found in 11 of the 14 microhabitats exploited by American lizards; see Appendix C). Among all 11 species of North American desert lizards, microhabitat niche breadth averages 2.19 (st. dev. = .999, $N$ = 11). In the southern hemisphere deserts, subterranean lizards add a fifteenth microhabitat resource state. Among 22 species of Kalahari lizards, observed microhabitat niche breadths vary from 1.0 in the very specialized fossorial *Typhlosaurus* skinks to nearly 6.0 (mean = 3.36, st. dev. = 1.44 $N$ = 22). Microhabitat generalists in the Kalahari include two species of climbing skinks (genus *Mabuya*) and the terrestrial lacertid *Ichnotropis squamulosa*. Mean microhabitat niche breadth is significantly greater in the Kalahari than it is in North America ($t = -.2.41$, $df = 31$, $P < .025$). In many of the 60 Australian species, microhabitat niche breadth is low (mean = 3.02, st. dev. = 1.38, $N$ = 60; see Appendix C), due partially to small sample sizes (Australian microhabitat niche breadths do not differ significantly from those in North America or in the Kalahari by *t*-tests). Numbers of lizards observed and their microhabitat niche breadths are, however, uncorrelated (Figure 3.1; $r$ = 0.037); many uncommon species, such as *Menetia greyi* and *Heteronotia binoei*, are nevertheless relatively generalized. Microhabitat niche breadths are, on average, broadest in the Kalahari, narrowest in North America, and intermediate in Australia.

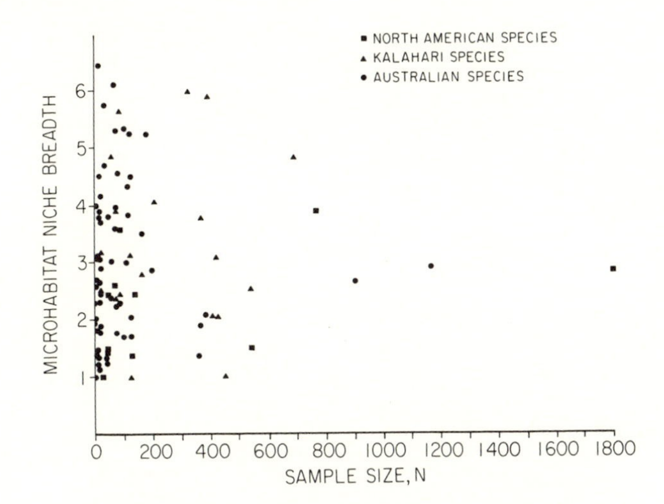


FIGURE 3.1. Contrary to popular belief, estimates of niche breadth can be independent of sampling effort provided that resource utilization patterns are well characterized for most species.

Table 3.2 gives estimates of the diversity of microhabitats actually used by the entire saurofaunas of each of the study areas. Note that microhabitat diversity is lowest in North America and most variable from site to site in the Kalahari. The diversity of microhabitats used by Australian desert lizards is high.

Thermal relations of active lizards, which vary widely among species, are profoundly influenced by spatial and temporal patterns of activity. Body temperatures of some diurnal heliothermic species average 38°C or higher, whereas those of nocturnal thigmothermic species are usually in the mid-twenties (Appendix D gives statistics for a variety of species).

Interesting interspecific differences also occur in the variance in body temperature as well as in the relationship between body temperatures and air temperatures. For example, among North American lizards, the two arboreal species (*Urosaurus graciosus* and *Sceloporus magister*) display narrower variances in body temperature than do terrestrial species. Presumably arboreal habits often facilitate efficient, economic, and rather precise thermoregulation. Climbing lizards have only to shift position slightly to be in the sun or shade or on a warmer or cooler substrate, and do not normally have to move through a diverse thermal environment. Moreover, arboreal lizards need not ex-

TABLE 3.2. Estimates of diversity of microhabitats used by the entire lizard faunas of various study sites (based on site-specific data).

| NORTH AMERICA | | KALAHARI | | AUSTRALIA | |
|---|---|---|---|---|---|
| *Site* | *Microhabitat Diversity* | *Site* | *Microhabitat Diversity* | *Site* | *Microhabitat Diversity* |
| I | 2.77 | A | 6.58 | A | 8.82 |
| L | 2.67 | B | 9.10 | D | 7.86 |
| G | 2.41 | D | 4.44 | E | 7.56 |
| V | 2.52 | G | 4.32 | G | 6.66 |
| S | 2.56 | K | 10.56 | $L_1$ | 6.36 |
| P | 2.18 | L | 5.93 | $L_2$ | 5.37 |
| M | 2.99 | M | 6.80 | M | 6.96 |
| T | 2.89 | R | 3.24 | N | 5.05 |
| W | 3.25 | T | 9.35 | R | 7.36 |
| C | 3.58 | X | 8.05 | Y | 8.06 |
| Mean | 2.78 | | 6.84 | | 7.01 |
| Standard deviation | 0.41 | | 2.42 | | 1.18 |

pend energy making long runs as most ground-dwellers must, and thus climbing species do not raise their body temperatures metabolically to as great an extent as terrestrial lizards.

Differences in temporal patterns of activity, the use of space, and body temperature relationships are not independent. Rather, they complexly constrain one another, sometimes in intricate and obscure ways. For example, thermal conditions associated with particular microhabitats change in characteristic ways in time: a choice basking site at one time of day becomes an inhospitable hot spot at another time (see also Schoener 1970). Perches of arboreal lizards receive full sun early and late in the day when ambient air temperatures tend to be low and basking is therefore desirable, but these same tree trunks are shady during the heat of midday when heat-avoidance behavior is more appropriate (Huey and Pianka 1977c). In contrast, the fraction of the ground's surface in the sun is low when shadows are long early and late but reaches a maximum at midday. Terrestrial heliothermic lizards may thus experience a shortage of suitable basking sites early and late in the day; moreover, during the heat of the day, their movements through relatively extensive patches of open sun can be severely curtailed. Hence ground-dwelling lizards encounter fundamentally different and more difficult thermal challenges than do climbing species.

During the warm summer months, many species of diurnal lizards display a bimodal daily pattern of activity, but, at cooler times of the year, activity is restricted to a single period during the midday (Figure 3.2). In other species, a unimodal period of activity changes gradually with the seasons (Figure 3.3). Such seasonal shifts in time of activity facilitate thermoregulation by allowing the animals to encounter a similar thermal environment at different times of year. Standardizing times of activity to "time since sunrise" for diurnal species or "time since sunset" for nocturnal ones helps to correct for such seasonality in activity times and facilitates comparisons among species and between communities. As indicated above, patterns of activity among sympatric lizard species often differ, with some emerging later after others have become less active. Such a sequential replacement of congeners during the day is illustrated in Figure 3.4 among five species of Australian *Ctenotus* skinks. *Ctenotus dux* emerges very early, followed by *C. calurus* (which has a narrow range in time of activity), then by *C. quattuordecimlineatus* and *C. piankai*, and finally during the heat of midday by *Ctenotus leae*. Body-temperature relationships reflect and presumably evolve in conjunction with these temporal differences in activity: mean body temperatures of these five species of *Ctenotus*, listed from earliest to latest, are 32.2°C, 36.0°C, 35.8°C,

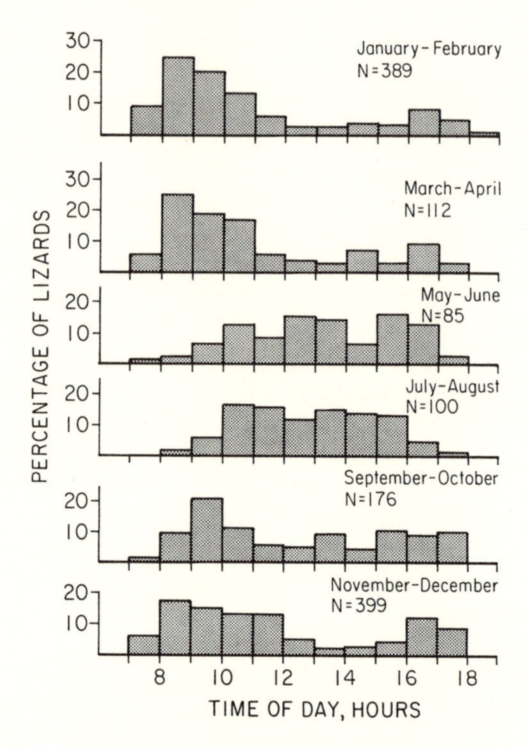


FIGURE 3.2. Frequency distributions of numbers of active *Ctenophorus isolepis* (an Australian desert agamid lizard) during different times of day at bimonthly intervals. Note the bimodal activity during the warmer months (November–April), but unimodal activity during the cooler months (especially July and August).

35.8°C, and 38.0°C, respectively (average body weights, in the same rank order, are 3.3, 1.4, 4.1, 1.5, and 3.1 gms.).

Similar sequential patterns of activity and correlations with body temperatures occur among North American lizards (Pianka 1973, 1977) and in Kalahari lacertids (Pianka et al. 1979). Differences in times of activity like these might be important since they could act to reduce the intensity of interspecific competition. Interference competition would be ameliorated by virtue of a lower frequency of contacts, whereas exploitation competition would be reduced if resources (such as food types or choice basking sites) are either renewed rapidly or encountered differentially.

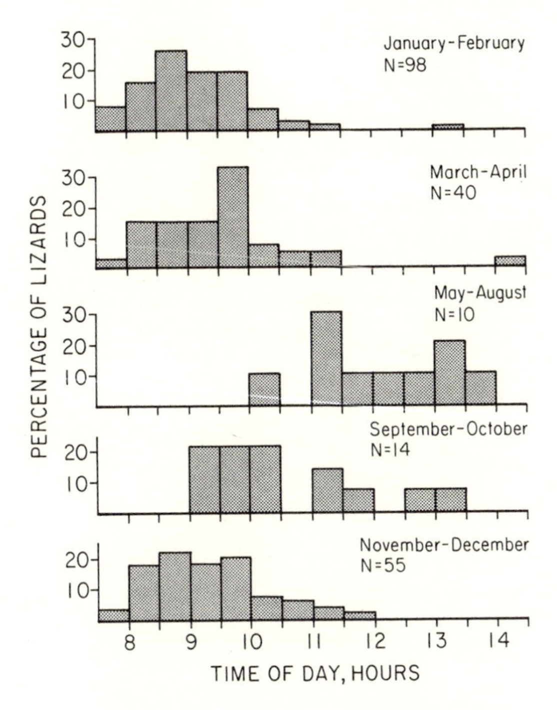


FIGURE 3.3. Frequency distributions of times of activity of 217 *Ctenotus calurus* at bimonthly intervals to show seasonal change. When expressed as times since sunrise, these data become more compact (see second panel of Figure 3.4).

Sample sizes of some common species are large enough to allow detailed analyses of body-temperature relationships. Among North American species, the whiptail *Cnemidophorus tigris* is abundant and found on all study areas, enabling some interesting geographic comparisons (Pianka 1970c). *Cnemidophorus* compensate both physiologically and behaviorally for latitudinal differences in daily temperature regimens. These teiid lizards display a cline in average body temperature from north to south along the North American deserts with lower body temperatures in the cooler northern deserts (Figure 3.5). *Cnemidophorus* and other lizards also emerge later in the day in the north, when ambient environmental temperatures are more comparable to those experienced by southern populations. Mean body temperature of Great Basin *Cnemidophorus* (38.9°C, $N$ = 293) is significantly lower than it is among southern whiptails (39.7°C, $N$ =

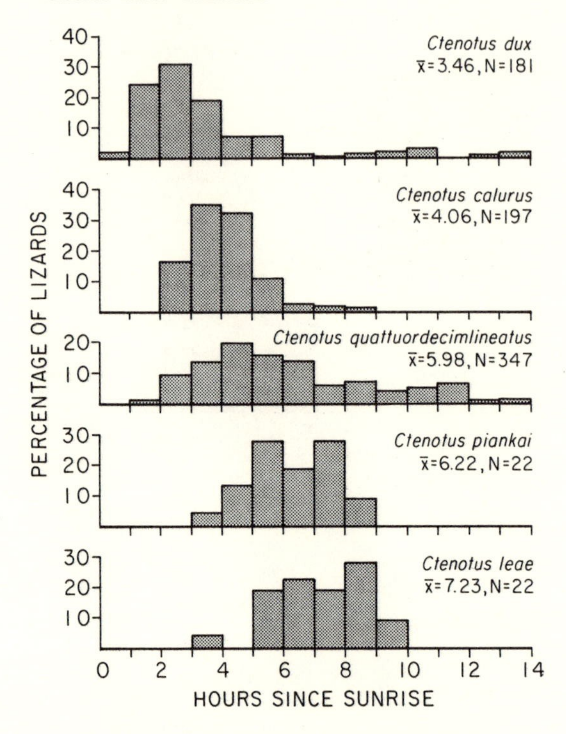


FIGURE 3.4. Times of activity among five species of Australian desert skinks of the genus *Ctenotus*, frequently found in sympatry.

1645). The difference in air temperature at the time of capture is even more pronounced: among 276 northern animals, average air temperature is only 24.8°C, whereas that for 1507 southern lizards is 27.5°C.

Another abundant species is the Australian agamid *Ctenophorus isolepis*. These long-legged lizards are denizens of the open spaces and make long zigzag runs between spinifex tussocks when disturbed. Body temperatures of *Ctenophorus isolepis* change seasonally (Figure 3.6) as well as during the day (Figure 3.7). In summer, the percentage of *C. isolepis* in the sun is very high early in the day, but it decreases rapidly as air temperatures rise (Figure 3.8). Moreover, lizards in the shade when first sighted actually have *higher* body temperatures (mean = 38.5°C, $N$ = 132) than do those that are in the sun (mean = 37.7°C, $N$ = 1129), strongly suggesting active shade seeking. This difference is highly significant ($P < 0.001$) using a $t$-test.

Fairly detailed analyses of temporal trends in body temperatures

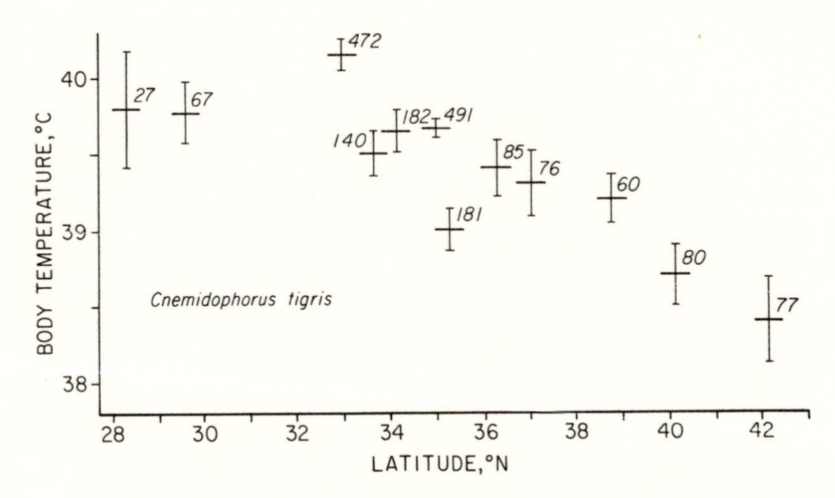


FIGURE 3.5. Body temperature changes with latitude in the teiid lizard, *Cnemidophorus tigris* (data from Pianka 1970c).

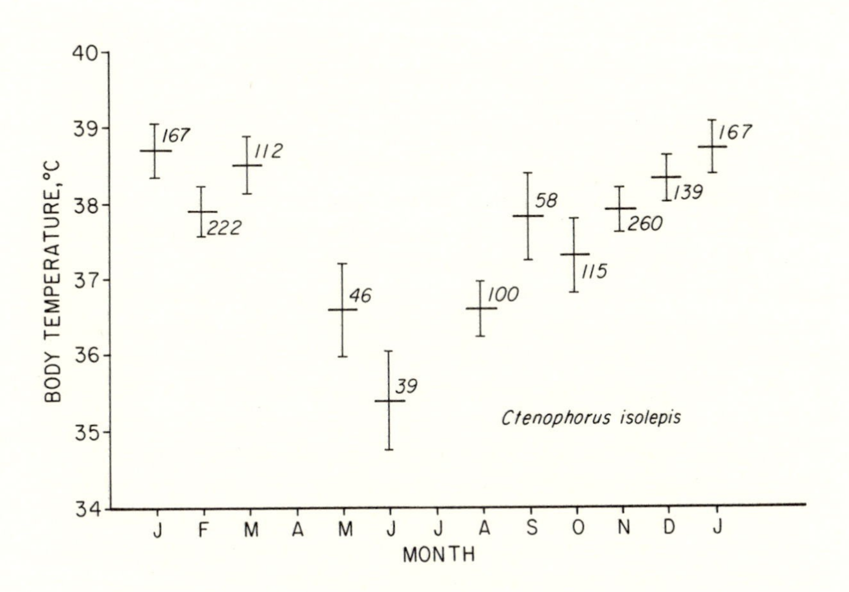


FIGURE 3.6. Seasonal changes in body temperatures of *Ctenophorus isolepis*.

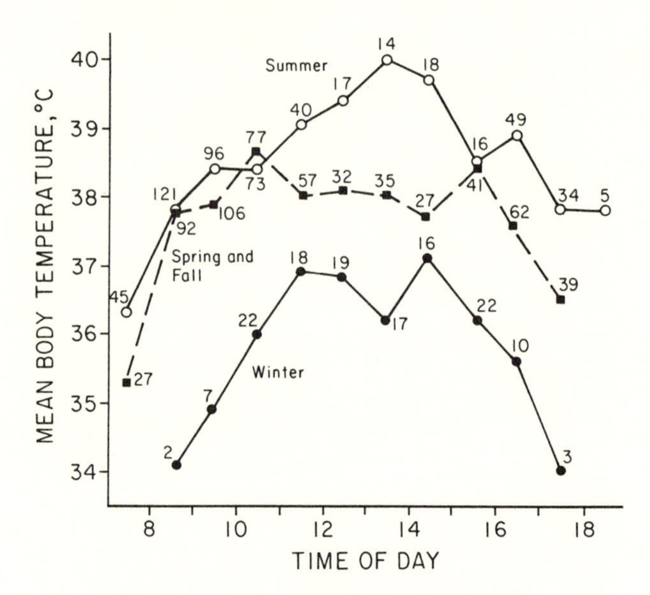


FIGURE 3.7. Average body temperatures of *Ctenophorus isolepis* at different times of day during different seasons. Summer = December, January and February; Winter = June, July, and August. Numbers represent sample sizes.

and thermoregulatory behavior among 12 species of diurnal Kalahari lizards (Huey and Pianka 1977c) show clear seasonal variation in times of activity, in body temperatures, and in the percentage of lizards in the sun. (The latter two statistics also vary within the day.) As in many lizards, body temperatures are correlated with air temperatures among these species, but the *difference* between the temperature of ambient air and that of active lizards tends to increase as air temperatures fall due to basking activities on the part of the lizards. Analogous patterns occur in North American and Australian desert lizards.

In a cost-benefit analysis of lizard thermoregulatory strategies, Huey and Slatkin (1976) identified the slope of the regression of body temperature against ambient environmental temperature as a useful indicator (in this case, an inverse measure) of the degree of passiveness in regulation of body temperature. On such a plot of active body temperature versus ambient air temperature, a slope of one indicates true poikilothermy or totally passive thermoconformity (a perfect correlation between air temperature and body temperature results), whereas a slope of zero reflects the other extreme of perfect thermoregulation. Lizards span nearly this entire thermoregulation spectrum:

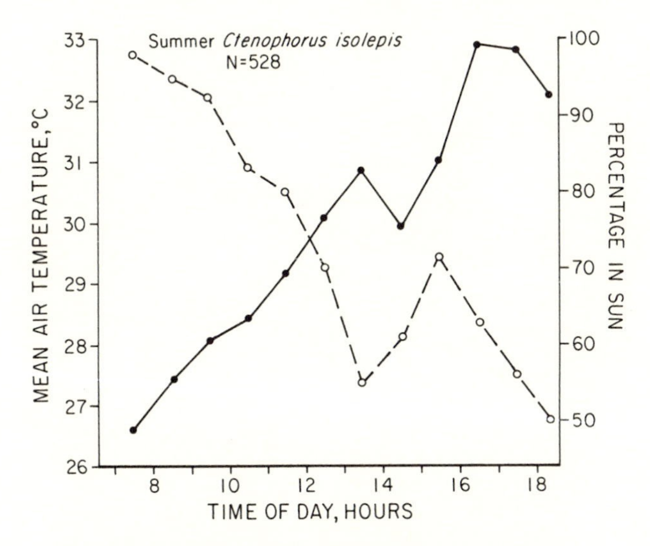


FIGURE 3.8. Percentages of *Ctenophorus isolepis* in the sun when first sighted decreases with time of day (dashed line), whereas average air temperature rises (solid line).

among active diurnal heliothermic species, regressions of body temperature on air temperature are usually fairly flat (for several species, slopes do not differ significantly from zero); among nocturnal species, slopes of similar plots are typically closer to unity (Appendix D). Various other species, particularly Australian ones, are intermediate, filling in this continuum of thermoregulatory tactics.

A straight line can be represented as a point in the coordinates of slope versus intercept. These two parameters are plotted for linear regressions of body temperatures on air temperatures among some 82 species of lizards, for which I have collected data, in Figure 3.9. Each data point represents the least-squares linear regression of body temperature against air temperature for a given species of desert lizard (see Appendix D for data). Interestingly enough, these data points fall on yet another, transcendent, straight line (no a priori reason for a linear dependence is evident; it is *not* a simple mass-related phenomenon). The position of any particular species along this spectrum reflects a great deal about its complex activities in space and time. The line plotted in Figure 3.9 thus offers a potent linear dimension on which various species can be placed in attempts to formulate general schemes of lizard ecology. It proves instructive to determine the extent to which various other ecological parameters, including foraging

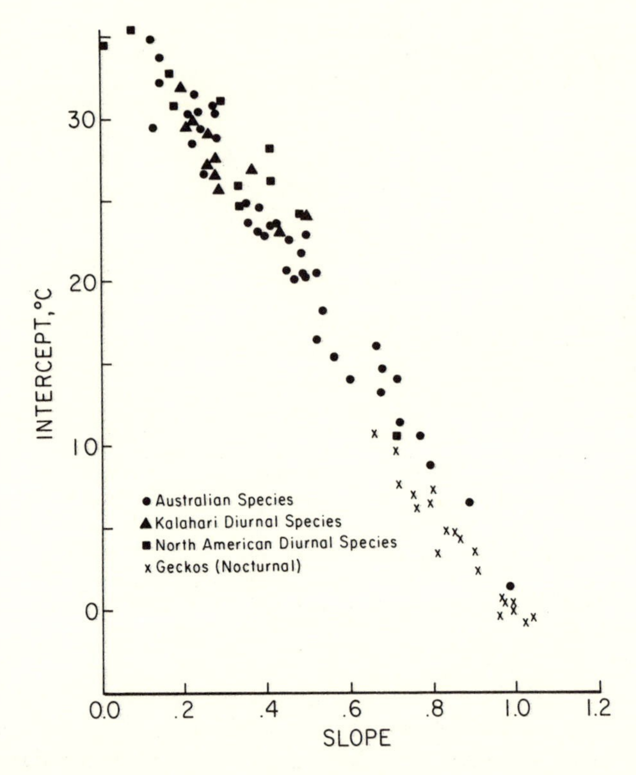


FIGURE 3.9. Each data point represents the least-squares linear regression of body temperature against air temperature for a given species of desert lizard (see Appendix D for statistics). Sample sizes are usually substantial (mean N is 145; data on a total of 11,924 individual lizards were used to make this plot). The ordinate represents the spectrum of thermoregulatory tactics ranging from active thermoregulators (slope = 0) to entirely passive thermoconformity (slope = 1). The intriguing "intercept" of the intercepts (38.8°C) approximates the point of intersection of all 82 regression lines and presumably represents an innate design constraint imposed by lizard physiology and metabolism. Even so, this statistic varies among taxa, ranging from only 29.4°C in 19 species of nocturnal geckos to 39.1°C in 13 species of diurnal agamids (in other diurnal groups, such as skinks and iguanids, it is about 36.5 to 37°C).

modes and reproductive tactics, can be mapped either directly onto or at right angles (orthogonally) to this emergent spatial-temporal axis (a "periodic table" of lizard niches could emerge from such considerations).

The intriguing "intercept" of the intercepts (38.8°C) approximates the point of intersection of all 82 regression lines and presumably represents an innate design constraint imposed by lizard physiology and metabolism. Moreover, this value also corresponds more or less to the body temperature of homeotherms, particularly mammals.

In a principal components analysis of these slopes and intercepts, the first principal component reduces variance by a full 99%. Microhabitat overlap among species is weakly but highly significantly inversely correlated with the distance between 79 species on this principal component ($r = -.32$, $P < .01$), confirming that this thermoregulatory axis does indeed reflect spatial patterns of activity.

# 4 Modes of Foraging and Trophic Relationships

Many predators attack their prey from ambush, but others usually hunt while on the move. These two modes of foraging have been called the "sit-and-wait" versus the "widely foraging" tactic, respectively (Pianka 1966; Schoener 1971). Of course, this dichotomy is somewhat artificial, although numerous animal groups seem to fall rather naturally into either one category or the other. Members of most lizard families typically exploit either one or the other of these two modes of foraging: thus iguanids, agamids, and geckos primarily sit and wait for their prey, whereas teiids and most skinks forage widely. Lacertids, however, use both modes of foraging, even within the same genus (see below).[1] This evidently natural dichotomy in foraging tactics has had a substantial impact on theories of optimal diets and competitive relationships among species. Yet empirical study of the influence of mode of foraging on diets has lagged behind theory.

Not unexpectedly, certain dietary differences are associated with this apparent dichotomy in foraging tactics. One would expect sit-and-wait predators to rely largely on moving prey, whereas widely foraging predators should encounter and consume nonmoving types of prey items more frequently. In order for the sit-and-wait tactic to pay off, prey must be relatively mobile and prey density must be high (or predator energy requirements low). One might predict, then, that the sit-and-wait tactic would be less prevalent during periods of prey scarcity than the widely foraging method. The success of the widely foraging tactic is also influenced by prey mobility and prey density as

[1] One Namib desert lacertid species, *Aporosaura anchietae*, has actually been reported to "switch" from sitting and waiting for windblown seeds to foraging widely for insect prey when winds are calm (Robinson 1978).

well as by the predator's energetic requirements (which should usually be higher than those of sit-and-wait predators), but the searching abilities of the predator and the spatial distribution of its prey now assume substantial importance.

North American and Australian sites support similar numbers of species of sit-and-wait foragers, whereas this mode of foraging is distinctly impoverished in the Kalahari (see Table 1.2). Markedly fewer species forage widely in western North America (only one species, the teiid *Cnemidophorus tigris*) and in the Kalahari (an average of 4 species per site) than in the Australian deserts (mean number of widely foraging species per area is 10.1, most of which are skinks in the genus *Ctenotus*). Intercontinental comparisons of proportions of total species in various foraging modes are also instructive: a full 60% of North American lizard species are sit-and-wait foragers, compared to only 16% in the Kalahari and 18% in Australia; percentages of widely foraging species are 14% (North America), 27% (Kalahari), and 36% (Australia).

Two species of Kalahari lizards, *Eremias lineo-ocellata* and *Meroles suborbitalis*, sit-and-wait for prey, whereas several other syntopic species, including two other species in the genus *Eremias* (*E. lugubris* and *E. namaquensis*), forage widely for their food (Pianka et al. 1979; Huey and Pianka 1981). Time budgets of these lacertids reflect their modes of foraging (Table 4.1 and Figure 4.1). Foraging widely is energetically expensive and, judging from their relative stomach volumes, those species that engage in this mode of food gathering appear to capture more prey per unit time than do sit-and-wait species. Indeed, overall energy budgets of widely foraging species are approximately half again as great as those of sit-and-wait species (Huey and Pianka 1981). As would be expected, sedentary foragers tend to encounter and eat fairly mobile prey, whereas more active widely foraging predators consume less active prey. Compared with sit-and-wait species, widely foraging lacertid species eat more termites (sedentary, spatially and temporally unpredictable but clumped prey). One widely foraging species, *Nucras tessellata*, specializes on scorpions. (By day, these large arachnids are nonmobile and exceedingly patchily distributed prey items.)

Another ramification of foraging mode in these Kalahari lizards concerns exposure to their own predators. Because of their more or less continual movements, widely foraging species tend to be more visible and, as a result, seem to suffer higher predation rates. Widely foraging species fall prey to lizard predators that hunt by ambush, whereas sit-and-wait lizard species tend to be eaten by predators that

TABLE 4.1. Time-budget data on the foraging tactics of seven species of lacertid lizards in the Kalahari.

| SPECIES | N | Total Time Observed (Minutes) | Meters Moved per Minute | Distance per Move (Meters) | Number of Moves per Minute | Proportion of Time Spent Moving |
|---|---|---|---|---|---|---|
| *Eremias lineo-ocellata* | 15 | 152.5 | 1.23 ± 0.37 | 0.76 ± 0.07 | 1.54 ± 0.42 | 0.143 ± 0.030 |
| *Eremias lugubris* | 15 | 72.1 | 5.24 ± 0.40 | 1.93 ± 0.19 | 2.97 ± 0.28 | 0.574 ± 0.038 |
| *Eremias namaquensis* | 25 | 131.3 | 4.67 ± 0.52 | 1.79 ± 0.11 | 2.78 ± 0.31 | 0.535 ± 0.052 |
| *Ichnotropis squamulosa* | 5 | 20.9 | 3.20 ± 0.41 | 1.12 ± 0.18 | 3.10 ± 0.14 | 0.546 ± 0.079 |
| *Meroles suborbitalis* | 15 | 122.8 | 1.02 ± 0.14 | 0.58 ± 0.06 | 1.83 ± 0.19 | 0.135 ± 0.016 |
| *Nucras tessellata* | 11 | 59.7 | 7.71 ± 1.23 | 2.23 ± 0.33 | 2.90 ± 0.37 | 0.502 ± 0.052 |
| *Nucras intertexta* | 3 | 8.9 | 8.93 ± 2.46 | 2.53 ± 0.87 | 3.69 ± 0.27 | 0.649 ± 0.039 |

NOTE: Means, plus or minus one standard error.

forage widely, generating "crossovers" in foraging mode between trophic levels. Widely foraging lizard species are also more streamlined and have longer tails than sit-and-wait species (Huey and Pianka 1981).

Yet another spinoff of foraging mode involves reproductive tactics. Clutch sizes of widely foraging species tend to be smaller than those of sit-and-wait species (see also next chapter), probably because the former simply cannot afford to weight themselves down with eggs to as great an extent as can the latter. Hence foraging style constrains reproductive prospects (as well as vice versa). Huey and Pianka (1981) summarize some of these ecological correlates of foraging mode. Similar patterns have been described in certain insectivorous birds by Eckhardt (1979).

Certain species of lizards are dietary specialists, eating only a very narrow range of prey items. For example, the Australian agamid *Moloch horridus* eats essentially nothing except ants, mostly of a single species of *Iridomyrmex* (North American horned lizards, genus *Phrynosoma*, are also ant specialists; see also below). Other species are termite specialists, including the Kalahari lizards *Eremias lugubris* and *Typhlosaurus* and the Australian nocturnal *Diplodactylus conspicillatus, Rhynchoedura*, as well as some diurnal *Ctenotus* species. Even though these species eat virtually nothing but isoptera, other species

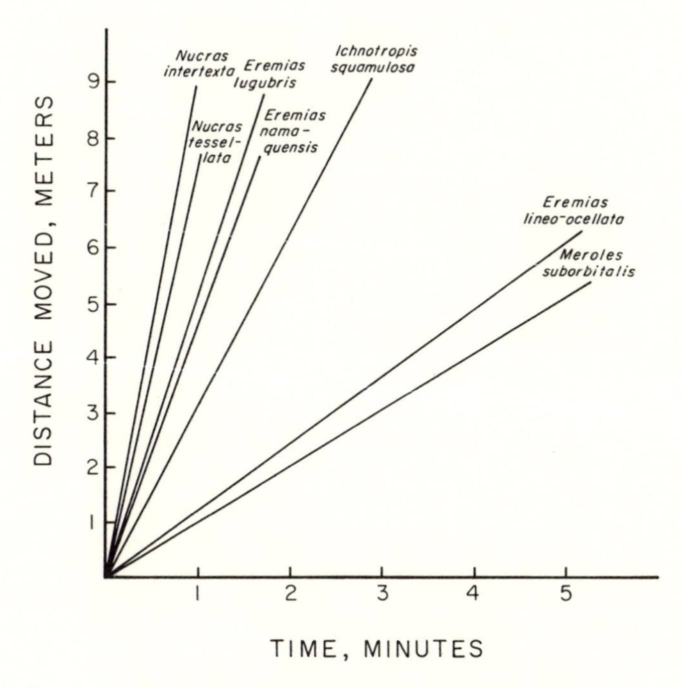


FIGURE 4.1. Average rates of movement among seven species of lacertid lizards in the Kalahari.

never touch termites as prey (see Appendix E). Food specialization on termites and ants is economically feasible because these social insects normally occur in a clumped spatial distribution and hence constitute a concentrated food supply. Still other lizard species, while not quite so specialized, also have narrow diets. For example, the Kalahari lacertid *Nucras tessellata* (mentioned above) and the Australian pygopodid *Pygopus nigriceps* both consume an excessive number of scorpions compared to other lizard species. *Nucras* forages widely to capture these large arachnids by day in their diurnal retreats, whereas the nocturnal *Pygopus* sits and waits for scorpions at night above ground during the latter's normal period of activity.[2] While scorpions are solitary prey items, they are extremely large and nutritious, presumably facilitating evolution of dietary specialization. For similar reasons, specialization on other lizards as food items has evolved in

[2] Interestingly, no North American desert lizard has evolved into a scorpion specialist, even though these large arachnids are moderatley plentiful. Perhaps the small snake *Chionactis occipitalis* has usurped this ecological role (Norris and Kavanau 1966).

North American leopard lizards (*Crotaphytus wislizeni*) as well as among most Australian varanids (see below). Other lizard species have much more catholic diets and eat a considerably wider variety of foods. Dietary diversity also varies within species from time to time and from place to place as the composition of diets changes with opportunistic feeding in response to fluctuating prey abundances and availabilities (Figures 4.2 and 4.3). However, the *consistency* of lizard diets is fairly remarkable (see also Chapter 10), suggesting a profound impact of microhabitat utilization, foraging mode, as well as various anatomical and behavioral constraints.

In an environment with a scant food supply, a consumer presumably cannot afford to bypass many inferior prey items, because mean search time per item encountered is long and expectation of prey encounter is low (MacArthur and Pianka 1966). In such an environment, a broad diet maximizes returns per unit expenditure, favoring generalization. In a food-rich environment, however, search time per item is low since a foraging animal encounters numerous potential prey items; under such circumstances, substandard prey items can be bypassed economically because expectation of finding a superior item in the near future

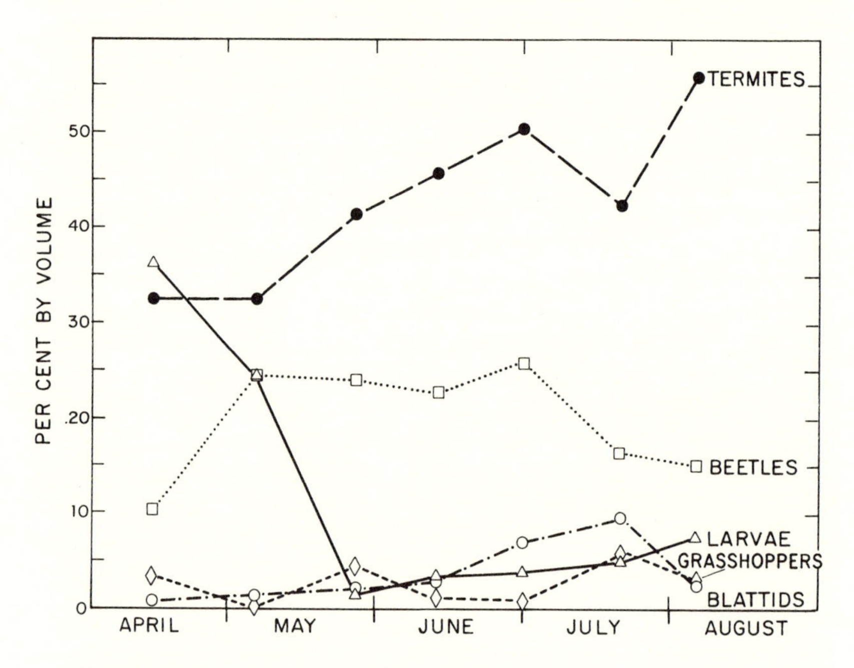


FIGURE 4.2. Seasonal changes in diet in the North American teiid *Cnemidophorus tigris* at a Sonoran desert study site (area C).

1. Tracks of Australia's largest living species of lizard, the perentie *Varanus giganteus*, impress the sands of a dune at "Red Sands," one of the author's favorite study areas in the Great Victoria desert of remote interior Western Australia. Achieving lengths in excess of two meters, perenties are exceedingly intelligent and wary lizards, ecologically equivalent to mammalian carnivores on other continents. A Pleistocene fossil relative, *Varanus prisca*, is estimated to have been nearly ten meters long!

2. The Kalahari lacertid *Nucras tesselata*, a widely foraging lizard active during the heat of midday at very high ambient temperatures. These lizards are scorpion specialists, digging up these large arachnids in their diurnal retreats.

3. The chameleon *Chamaeleo dilepis*, an uncommon arboreal lizard encountered rarely in the Kalahari. This bizarre lizard family is characterized by its peculiarly split feet, with a two-three arrangement of opposing digits. Their tongues are as long as the lizard and are used to ambush insects in a sit-and-wait mode of foraging.

4. An uncommon arboreal gecko, *Pachydactylus rugosus*, found on a variety of small woody shrubs and trees in the Kalahari.

5. A juvenile Kalahari lacertid *Eremias lugubris* that mimics a noxious carabid beetle known as an "oogpister" (euphemistically translatable as "eye squirter"). These beetles emit a pungent mixture of aldehydes, acids, and other noxious chemicals. Baby lizards are jet black and exhibit a most peculiar "stilt walk," stiff legged with their backs arched dorsal-ventrally and their reddish tails dragging the red Kalahari sand. As the lizards approach the size of the largest oogpister beetles, they abandon the stilt walk and metamorphose into a paler beige adult coloration.

6. A Kalahari barking gecko, *Ptenopus garrulus*, normally nocturnal, was active on an overcast day eating swarming winged alate termites. Local shrikes had a field day catching and crucifying these lizards.

7. An adult male Kalahari agamid, *Agama hispida*, in breeding color; these lizards climb and eat a lot of ants.

8. An Australian agamid, *Gemmatophora longirostris* (formerly assigned to *Physignathus* and *Lophognathus*); in the Great Victoria desert, this climbing species favors bloodwood trees and is usually found in close association with sandridges.

9. Another, considerably smaller, Australian agamid, *Diphoriphora winneckei*, found in sympatry with *Gemmatophora*. These graceful acrobats frequent large annuals and shrubs on sandridges in the Great Victoria desert. Diets of insectivorous climbing lizards such as these tend to show high electivities for certain types of flying and/or climbing insects, such as wasps, flies, mantids and phasmids, hemipterans and homopterans.

10. *Ctenophorus clayi* (formerly *Amphibolurus*), an Australian agamid found in sandridge habitats of the Great Victoria desert.

11. *Ctenophorus reticulatus* (formerly *Amphibolurus*), an agamid found in *Acacia*-shrub habitats in central Australia.

12. The "mountain devil," *Moloch horridus*, a bizarre agamid lizard found throughout most of the arid Australian interior. These lizards exhibit an extreme degree of dietary specialization, consuming virtually nothing but ants.

13. *Ctenophorus fordi* (formerly *Amphibolurus*), a sandridge species of agamid in the Great Victoria desert of Western Australia.

14. The "race horse goanna," *Varanus tristis*, an arboreal Australian varanid that preys upon other lizards, large insects, and baby birds, particularly parrots. These large lizards are inactive for months on end and leave a very distinctive track.

15. The perentie *Varanus giganteus*, Australia's largest species of lizard, formerly thought to require rock outcrops but now known to occur in sandy desert habitats as well. Like all varanids, perenties are exceptionally wary lizards, difficult to observe directly. Fortunately, however, one can deduce a lot about their activities and behavior from careful study of their tracks.

16. An arboreal pygmy goanna, *Varanus gilleni*. Uncommon in the Great Victoria desert, these small varanids prey largely on other climbing lizards, particularly on the gecko *Gehyra*.

17. The terrestrial pygmy monitor, *Varanus eremius*, found ubiquitously on all study sites in the Great Victoria desert. These searching predators, the author's favorite species of lizard, forage over vast distances and subsist almost exclusively on other species of desert lizards.

18. An *Omolepida branchialis* skink at the edge of a *Triodia* tussock, its favored microhabitat.

19. A worm-like subterranean Australian skink, *Lerista desertorum*, that feeds on small insects, especially termites. In the Kalahari semidesert of southern Africa, the skink *Typhlosaurus* is crudely ecologically equivalent.

20. The desert blue-tongued lizard, *Tiliqua multifasciata*, an uncommon, partially herbivorous Australian skink.

21. An elusive spinifex dweller, this small diurnal skink (*Ctenotus piankai*) was one of seven that had not yet been named when the author "discovered" them in 1966–68.

22. A gravid female leopard skink, *Ctenotus pantherinus*; these Australian desert lizards typically stay close to spinifex grass tussocks and feed heavily on termites.

23. *Ctenotus helenae*, a diurnal Australian skink typically found in messy spinifex-litter habitats beneath desert *Eucalyptus* trees.

24. A spinifex gecko, *Diplodactylus elderi*, found in *Triodia* habitat throughout most of arid interior Australia. These nocturnal lizards spend most of their lives inside grass tussocks, eating a variety of small insects.

25. An insectivorous Australian legless lizard, the pygopodid *Delma fraseri*, which frequents *Triodia* grass tussocks. Other Australian pygopodids are larger and eat other lizards and large scorpions.

26. This Australian knob-tailed gecko, *Nephrurus laevissimus*, is engaged in a threat display, hissing and lunging at its captors. This species occurs only in close proximity to sandridges; other species of *Nephrurus* are found on sandplains, on harder soils, and in association with rocks.

27. Sporting a satanic smile, this climbing Australian *Diplodactylus ciliaris* was located by its eyeshine using a head lantern. Capture is usually a relatively simple matter, except for the fact that these geckos exude a noxious sticky mucous from their glandular tails.

28. A close-up of the head and foreparts of the Australian agamid *Moloch*, showing the false head and sharp spines.

29. *Diplodactylus damaeus*, a little-known uncommon terrestrial Australian desert gecko.

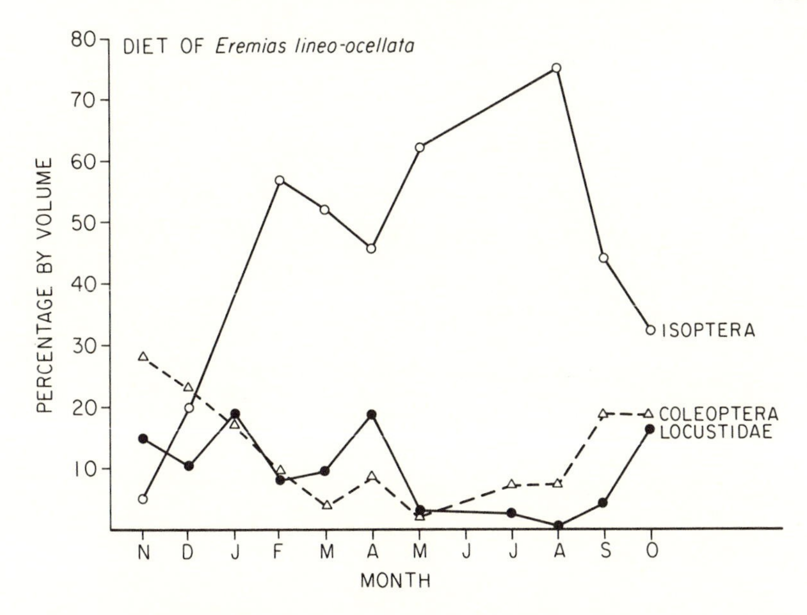


FIGURE 4.3. Seasonal changes in diet in the Kalahari lacertid *Eremias lineo-ocellata.*

is high. Hence rich food supplies are expected to favor selective foraging and to lead to narrow food-niche breadths. These arguments are supported by the North American teiid lizard *Cnemidophorus tigris*, which eats a greater diversity of foods in drier-than-average years (presumably times of low food availability) but like most lizards contracts its diet during periods of prey abundance (Figure 4.4). Another, more extreme, example of this phenomenon occurs after heavy summer rains when termites send out their winged alates in great abundance and virtually every species of lizard eats nothing but termites (even lizard species that normally never consume termites). During such fleeting moments of great prey abundance, there is little if any competition for food, and dietary overlap among members of a desert saurofauna is sometimes nearly complete.[3]

Let us now consider one of these dietary specialists in somewhat greater detail. North American desert horned lizards, *Phrynosoma platyrhinos*, as mentioned above, are ant specialists, eating little else. Various features of their anatomy, behavior, diet, temporal pattern of activity, thermoregulation, and reproductive tactics can be profitably

[3] Due to their very short duration, these bursts of high food availability with their associated extensive dietary overlap have only a trivial impact on *overall* utilization patterns such as those summarized in Appendix E.

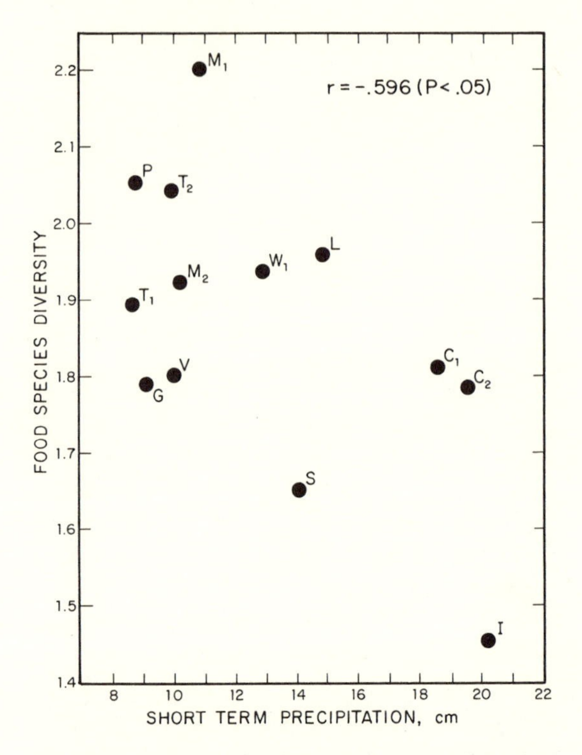


FIGURE 4.4. The diversity of prey eaten by *Cnemidophorus tigris* varies inversely with recent precipitation (food niche breadth contracts during wet, presumably productive, years).

interrelated and interpreted to provide an integrated view of the ecology of this interesting lizard (Pianka and Parker 1975). Ants are small and contain much undigestible chitin, so that large numbers of them must be consumed. Hence an ant specialist must possess a large stomach for its body size: the stomach of this horned lizard averages about 13% of the animal's overall body mass, a substantially larger fraction than stomachs of other lizard species, even among herbivorous lizards such as *Dipsosaurus* (in 6 other sympatric North American lizard species, stomach volumes average only 6.4% of body weight; among 8 Australian species selected as crude ecological counterparts, the ratio of stomach volume/body weight averages only 5.9%). *Phrynosoma*'s large gut requires a tank-like body form, reducing speed and decreasing the lizard's ability to escape from predators by movement. As a result, natural selection has favored a spiny body form and cryptic behavior rather than a sleek body and rapid movement to cover, as in the majority of other lizards. Long periods of exposure while foraging in the open presumably increase risks of predation. A reluctance to move, even when actually threatened by a potential predator, could well prove advantageous under such circumstances: movement might attract the predator's attention and negate the advantage of concealing coloration and contour. Such decreased movement doubtlessly contributes to the observed high variability in body temperature of *Phrynosoma platyrhinos*, which is significantly greater than that of all other sympatric species of lizards. Wide fluctuations in horned-lizard body temperatures under natural conditions presumably reflect both their long activity period and perhaps their reduced movements into or out of the sun and shade (when first encountered, most *Phrynosoma* are in the open sun). A consequence is that more time is made available for activities such as eating (foraging ant eaters must spend considerable time feeding). To make use of this patchy and spatially concentrated, but at the same time not overly nutritious, food supply, *P. platyrhinos* has evolved a unique constellation of adaptations that include its large stomach, spiny, tank-like body form, expanded period of activity, and "relaxed" thermoregulation (eurythermy). A further spinoff of the *Phrynosoma* adaptive suite concerns their extraordinarily high investment in reproduction. Females of some species devote as much as 35% of their body weight to production of a very large clutch of eggs (Pianka and Parker 1975). This is presumably a simple and direct consequence of their robust body form: lizards that must be able to move rapidly to escape predators would hardly be expected to weight themselves down with eggs to the same extent as animals such as horned lizards that rely almost entirely upon spines and camouflage to avoid predators.

Best estimates of the diets of 92 species of desert lizards are given in Appendix E: entries represent the percentages by volume of various foods in the overall diets of all specimens of each species on all study areas within each desert-lizard system. Close scrutiny of these data matrices reveals considerable texture: some species specialize on scorpions, ants, termites, vertebrates, and/or on plants, whereas other species on each continent are much more generalized, eating a wide variety of food categories. No centipedes were found in the diet of North American lizards, and solpugids are not present in Australia. Food niche breadths[4] range from 1.06 to 7.33 (mean = 4.40, st. dev.

[4] Computed using proportional utilization coefficients with Simpson's (1949) index of diversity.

= 2.01, $N$ = 11) among the 11 species of North American lizards, from 1.07 to 8.22 (mean = 3.89, st. dev. = 2.09, $N$ = 21) among 21 Kalahari species, and from 1.00 to 9.43 (mean = 3.81, st. dev. = 2.21, $N$ = 60) among the 60 Australian species. None of these intercontinental variations in food niche breadths are statistically significant by $t$-tests. Estimates of food niche breadths are uncorrelated with the number of lizards on which they are based (Figure 4.5; $r$ = 0.11), providing evidence that sample sizes are adequate to characterize patterns of food utilization even among the rarer species. Indeed, species with broad diets are often, though not always, relatively uncommon. Across species, dietary niche breadth is weakly, but significantly ($r = .268$, $P < .02$), positively correlated with microhabitat niche breadth, an indication that food specialists tend to be restricted to fewer microhabitats than food generalists.

Biologically significant variation occurs between species in utilization of certain relatively minor food categories: for example, in the diets of climbing lizard species, hemiptera-homoptera and mantids-phasmids as well as various flying insects (wasps, Diptera, and Lepidoptera) tend to be better represented than they are among terrestrial species. Likewise, geckos tend to consume more nocturnal arthropods (scorpions, crickets, roaches, and moths) than do most diurnal species (although certain diurnal lizards capture nocturnal prey in their diurnal

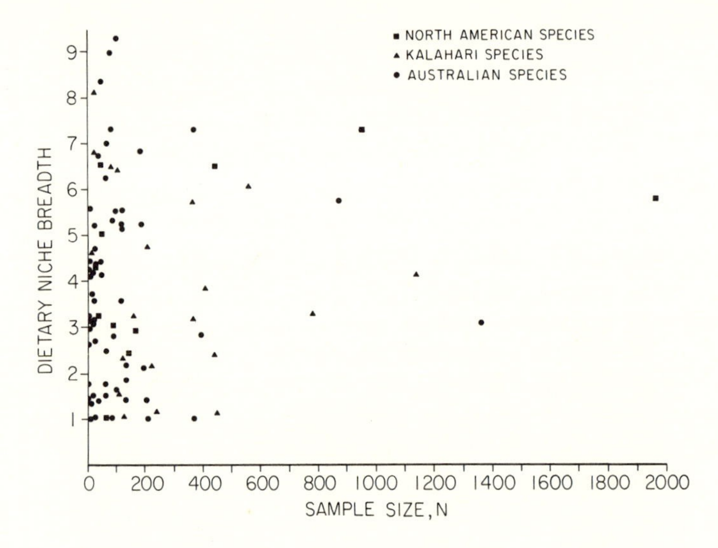


FIGURE 4.5. Estimates of dietary niche breadth are independent of sampling effort.

retreats; see Huey and Pianka 1983). Such prey items are thus indicators of spatial and temporal patterns of activity.

Overall diets of entire saurofaunas are summarized and compared in Table 4.2. Only a relatively few foods dominate these lizard diets. Prey resource spectra (Figure 4.6) are broadly similar between continents, although notable quantitative differences occur. In North America the seven most important categories (total 84%), in decreasing order by volumetric importance, are: beetles, termites, insect larvae, grasshoppers plus crickets, ants, plant materials, and vertebrates.

TABLE 4.2. Major prey items of entire saurofaunas.

| | NORTH AMERICA | | KALAHARI | | AUSTRALIA | |
|---|---|---|---|---|---|---|
| PREY CATEGORY | *Volume** | % | *Volume** | % | *Volume** | % |
| Centipedes | 0.0 | 0.0 | 0.0 | 0.0 | 56.2 | 2.0 |
| Spiders | 50.4 | 1.6 | 35.7 | 3.1 | 76.2 | 2.7 |
| Scorpions | 23.2 | 0.7 | 32.7 | 2.9 | 41.6 | 1.5 |
| Solpugids | 45.7 | 1.4 | 17.6 | 1.5 | — | — |
| Ants | 307.4 | 9.7 | 155.3 | 13.6 | 467.5 | 16.8 |
| Wasps | 27.6 | 0.9 | 8.7 | 0.8 | 38.1 | 1.4 |
| Locustidae | 363.8 | 11.5 | 70.0 | 6.1 | 239.5 | 8.6 |
| Blattidae | 100.0 | 3.2 | 4.2 | 0.4 | 77.9 | 2.8 |
| Mantids-Phasmids | 24.5 | 0.8 | 1.2 | 0.1 | 15.8 | 0.6 |
| Neuroptera | 7.0 | 0.2 | 0.5 | 0.04 | 1.9 | 0.1 |
| Coleoptera | 587.2 | 18.5 | 186.7 | 16.3 | 159.7 | 5.7 |
| Isoptera | 525.0 | 16.5 | 473.3 | 41.3 | 515.5 | 18.5 |
| Homoptera-Hemiptera | 30.6 | 1.0 | 15.2 | 1.3 | 48.4 | 1.7 |
| Diptera | 27.4 | 0.9 | 6.8 | 0.6 | 7.8 | 0.3 |
| Lepidoptera | 67.9 | 2.1 | 16.3 | 1.4 | 16.2 | 0.6 |
| Insect eggs | 10.0 | 0.3 | 0.2 | 0.02 | 1.7 | 0.1 |
| All larvae | 384.1 | 12.1 | 41.2 | 3.6 | 92.7 | 3.3 |
| Miscellaneous unidentified Arthropods | 83.5 | 2.6 | 40.0 | 3.5 | 58.4 | 2.1 |
| Vertebrates | 245.9 | 7.8 | 26.3 | 2.3 | 753.9 | 27.1 |
| Plants | 262.4 | 8.3 | 13.2 | 1.2 | 118.1 | 4.2 |
| Total volume* | 3173.6 | | 1145.1 | | 2787.1 | |
| Diversity of foods consumed by all species | 8.7 | | 4.4 | | 6.6 | |

* In cubic centimeters.

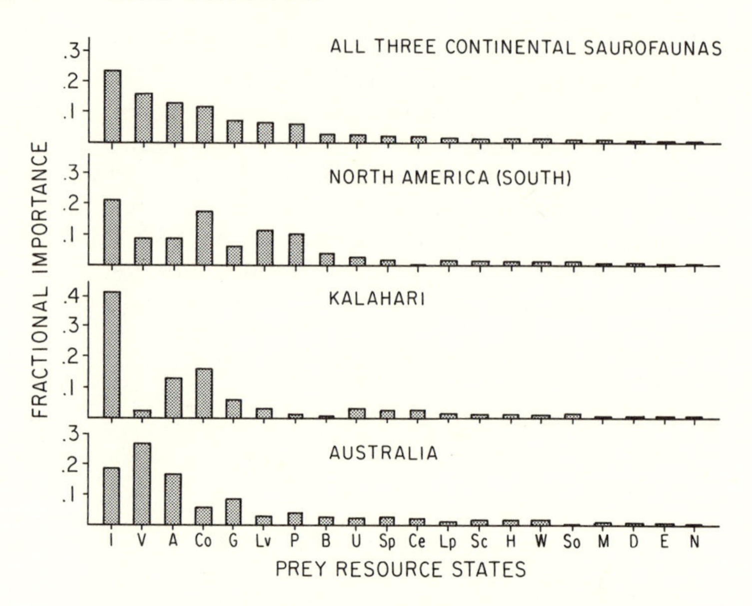


FIGURE 4.6. Comparison of prey resource spectra in each of the three continental desert-lizard systems. The panel at the top represents the overall diet of all lizards on all three continents. In this panel, prey types are ranked from those most used to those least used, based on prey volume. The ranking generated by the top panel is preserved in the lower three panels, which represent the actual diet of each continental saurofauna. Note that termites (I), beetles (Co), and ants (A) constitute major prey in all three systems.

In the Kalahari, just three food categories far outweigh all others (total 71%): termites, beetles, and ants. In Australia the five most important categories (total 77%), in decreasing order, are: vertebrates, termites, ants, grasshoppers plus crickets, and beetles. Note that the same three categories—termites, beetles, and ants—constitute major prey items in all three continental desert-lizard systems. Termites assume a disproportionate role in the Kalahari, as do vertebrate foods in Australia (largely a reflection of varanid diets). Somewhat surprisingly, the overall diversity of foods consumed by all species of lizards is actually greatest in the least diverse North American saurofauna (8.7), lowest in the Kalahari lizards (4.4), and intermediate in Australia (6.6). Basically comparable figures, although broadly overlapping, emerge from an area-by-area analysis (Table 4.3).

Estimates of prey diversity (Table 4.3) correlate weakly with certain measures of the variability in average annual precipitation: food diversity is positively correlated with the coefficient of variation in an-

TABLE 4.3. Estimates of diversity of foods eaten by the entire lizard faunas at various study sites (based upon diet by volume).

| NORTH AMERICA | | KALAHARI | | AUSTRALIA | |
|---|---|---|---|---|---|
| *Site* | *Food Diversity* | *Site* | *Food Diversity* | *Site* | *Food Diversity* |
| I | 4.25 | A | 5.64 | A | 7.39 |
| L | 4.17 | B | 5.24 | D | 7.65 |
| G | 6.54 | D | 5.45 | E | 3.74 |
| V | 6.01 | G | 5.60 | G | 4.66 |
| S | 4.75 | K | 4.15 | $L_1$ | 3.48 |
| P | 6.61 | L | 2.44 | $L_2$ | 4.28 |
| M | 7.32 | M | 5.97 | M | 6.50 |
| T | 8.50 | R | 1.93 | N | 7.86 |
| W | 6.51 | T | 2.77 | R | 5.19 |
| C | 6.64 | X | 5.64 | Y | 6.14 |
| Mean | 6.13 | | 4.48 | | 5.69 |
| Standard deviation | 1.38 | | 1.54 | | 1.64 |

nual precipitation ($r = .45, P < .05$), and is negatively correlated with the mean minus the standard deviation in precipitation ($r = -.49, P < .05$), Brown's (1975) measure of "the predictable amount of precipitation." Thus more variable precipitation, and presumably primary productivity, seems to foster higher insect species diversities (recall that lizard diversity also correlates with variability of precipitation; see page 25).

# 5 Reproductive Tactics

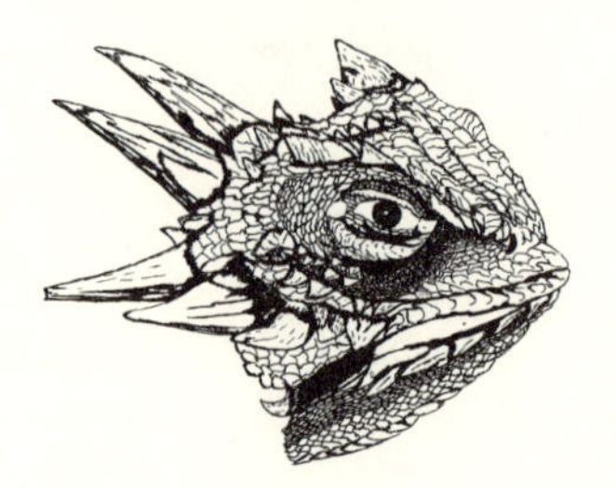

Most lizards lay eggs, but some species retain their eggs internally and give birth to living young. Clutch size varies from one to forty or more among different species of lizards. Some species reproduce only once every second or third year, others but once each year, while still others lay two or more clutches each year. Substantial spatial and temporal variation in clutch size also exists *within* species. Clutch size is fixed at one or two eggs in certain families (Geckos, Pygopodids) and genera (*Anolis*). Lizard reproductive tactics have attracted considerable attention (Ballinger 1983; Tinkle 1969; Tinkle et al. 1970; Tinkle and Hadley 1975; Vitt and Congdon 1978; Vitt and Price 1982).

As one of many possible examples, in the double-clutched Australian agamid species *Ctenophorus isolepis* (Pianka 1971c), the size of 67 first clutches (August–December) averaged 3.01 eggs, whereas the mean of 41 second clutches (January–February) was 3.88. Females increase in size during the season, and, as in many lizards, larger females tend to lay larger clutches. Interestingly enough, however, these same females appear to be investing relatively more on their second clutches than they do on their first: among 25 first clutches, clutch volumes average only 11.2% of female weight, but in 15 second clutches the average is 15.1%. (95% confidence intervals on these means are nonoverlapping—10.25 to 12.20 versus 13.38 to 16.85, respectively, Pianka and Parker 1975).

Changes in fecundity with fluctuations in food supplies and local conditions from year to year or location to location have also frequently been observed: for example, in the North American whiptail *Cnemidophorus tigris* (family Teiidae) females tend to lay larger clutches in years with above-average precipitation and presumably ample food supplies (Figure 5.1). Similar phenomena have been documented in the side-blotched lizard *Uta stansburiana* and doubtlessly occur in many other lizard species (Hoddenbach and Turner 1968; Parker and Pianka 1975; Dunham 1980).

Clutch or litter weight (or volume), expressed as a fraction of a female's total body weight, ranges from as little as 4 to 5% in some species to as much as 20 to 30% in others. Clutch weights tend to be particularly high in some of the North American horned lizards (genus *Phrynosoma*). Ratios of clutch or litter weight to female body weight correlate strongly with various energetic measures (Ballinger and Clark 1973; Vitt 1977) and have often been used as crude indices of a female's instantaneous investment in current reproduction (sometimes equated with the elusive notion of "reproductive effort").

In addition to clutch size and female total investment in reproduction, the size (or weight) of an individual oviducal egg or newborn progeny also varies widely among lizards, from as little as 1 to 2% in some species to a full 17% in the live-bearing Kalahari fossorial skink *Typhlosaurus gariepensis*. Such expenditures per progeny are inverse measures of the extent to which a juvenile lizard must grow to reach adulthood.

Of course, any two parties to this triad (clutch size, female reproductive investment, and expenditure per progeny) uniquely determine the third; however, forces of natural selection molding each differ substantially. Thus clutch or litter weight presumably reflects an adult

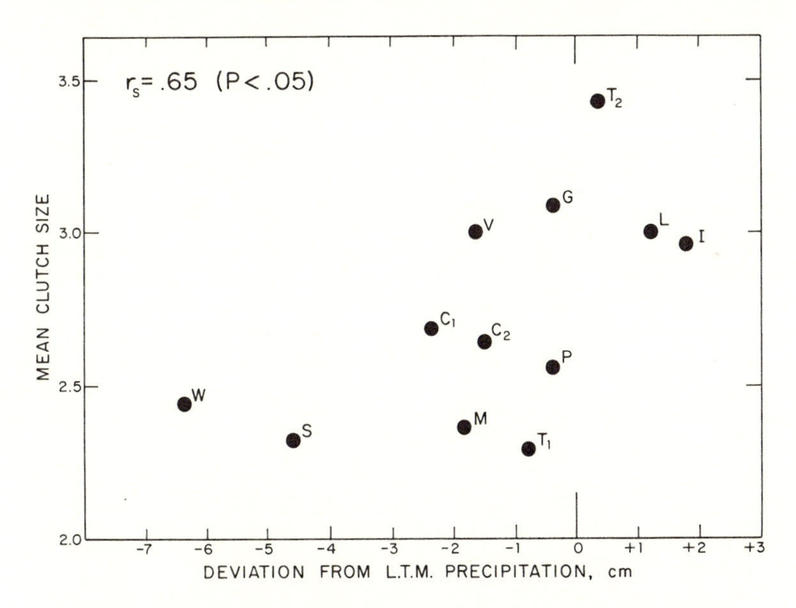


FIGURE 5.1. Average clutch size of female *Cnemidophorus tigris* plotted against the deviation of the short-term (last five years) precipitation from the long-term mean. Females tend to lay larger clutches in wetter years.

female's best current investment tactic in a given environment at a particular instant in time whereas expenditure on any given individual progeny is probably more closely attuned to the average environment to be encountered by a juvenile. In a sense, then, clutch (or litter) size is the direct result of the interaction between an optimal parental reproductive tactic and an optimal juvenile size (clutch size is, of course, simply the ratio of the former divided by the latter).

Statistics on clutch/litter sizes, total reproductive investment of females, and expenditure per progeny among 65 species of desert lizards are presented in Appendix F. For comparative purposes, similar data on another 20 species of lizards, including both desert and nondesert forms, were extracted from the literature (Table 5.1). Among the species surveyed, average clutch/litter size varies from 1 in the Kalahari skink *Typhlosaurus gariepensis* and the geckos *Gehyra variegata* and *Ptenopus garrulus* to 13 in the Kalahari agamid *Agama hispida*. Clutch

TABLE 5.1. Comparative data on reproductive tactics of some lizard species other than those listed in Appendix F.

| | CLUTCH/LITTER SIZE | | | REPRODUCTIVE EFFORT | | | |
|---|---|---|---|---|---|---|---|
| GENUS/SPECIES | $\overline{X}$ | SD | N | $\overline{X}$ | SD | N | AUTHORITY |
| *Cnemidophorus exsanguis* | 2.96 | 1.1 | 91 | 13.1 | 3.0 | 43 | Schall 1978 |
| *Cnemidophorus gularis* | 3.13 | 1.4 | 23 | 15.6 | 2.0 | 8 | Schall 1978 |
| *Cnemidophorus inornatus* | 2.37 | 0.92 | 87 | 14.7 | 4.0 | 28 | Schall 1978 |
| *Cnemidophorus tesselatus* | 3.23 | 1.1 | 56 | 12.6 | 4.0 | 11 | Schall 1978 |
| *Cnemidophorus tigris* | 2.02 | 0.7 | 43 | 10.5 | 4.0 | 12 | Schall 1978 |
| *Phrynosoma asio* | 16.9 | 4.9 | 7 | 35.4 | 1.7 | 2 | Pianka & Parker 1975 |
| *Phrynosoma braconnieri* | 8.4 | 2.3 | 5 | 21.4 | 2.8 | 3 | Pianka & Parker 1975 |
| *Phrynosoma cornutum* | 26.5 | 6.7 | 152 | 30.7 | — | 30 | Ballinger 1974 |
| *Phrynosoma cornutum* | 23.7 | 6.0 | 73 | 28.8 | 6.0 | 41 | Pianka & Parker 1975 |
| *Phrynosoma coronatum* | 12.6 | 3.8 | 69 | 21.3 | 3.0 | 16 | Pianka & Parker 1975 |
| *Phrynosoma douglassi* | 16.0 | — | 56 | 25.4 | 8.6 | 11 | Pianka & Parker 1975 |
| *Phrynosoma m'calli* | 5.4 | 1.3 | 121 | 27.0 | — | 1 | Pianka & Parker 1975 |
| *Phrynosoma modestum* | 10.6 | 2.3 | 40 | 25.8 | 5.9 | 18 | Pianka & Parker 1975 |
| *Phrynosoma orbiculare* | 9.4 | 3.6 | 26 | 12.8 | 2.5 | 15 | Pianka & Parker 1975 |
| *Phrynosoma solare* | 16.7 | 3.7 | 28 | 32.0 | 7.2 | 9 | Pianka & Parker 1975 |
| *Sceloporus jarrovi* | 10.5 | — | 52 | 32.6 | 4.9 | 21 | Ballinger 1971 |
| *Sceloporus malachiticus* | 6.4 | — | 54 | 34.4 | — | 7 | Marion & Sexton 1971 |
| *Sceloporus olivaceus* | 15.0 | — | — | 24.6 | 1.1 | 12 | Ballinger 1971 |
| *Sceloporus poinsetti* | 10.0 | — | 40 | 33.7 | 0.6 | 12 | Ballinger 1971 |
| *Sceloporus undulatus* | 7.6 | — | — | 26.1 | 0.4 | 14 | Ballinger 1971 |

sizes in certain horned lizards are still larger, averaging 24.3 in the American iguanid *Phrynosoma cornutum* (the Texas horned lizard). Clutch or litter size and female investment are significantly positively correlated ($r = .482$, $P < .001$), although scatter is considerable. Viviparous species (mostly skinks) tend to have slightly higher investment ratios than do the egg layers. Since expenditure per progeny can be estimated from total investment divided by clutch/litter size, it tends to decrease exponentially with increasing number of progeny (for a fixed total investment). Figure 5.2 shows the observed inverse relationship between expenditure per progeny and clutch/litter size ($r = -.652$, $P < .001$). This correlation is strengthened when both variables are transformed to logarithms (on a log-log plot, the correlation coefficient is $-.810$). Clutch size is thus correlated positively with total investment but negatively with investment per progeny. In simple product-moment correlation, the latter two members of the triumvirate—total reproductive investment and expenditure per progeny—are only weakly and not significantly correlated ($r = .153$, $P < .10$), suggesting that these two parameters vary independently of one another and that they may be responsive to different selective pressures. However, when effects of clutch size are held constant by partial correlation, the weak correlation between reproductive effort and ex-

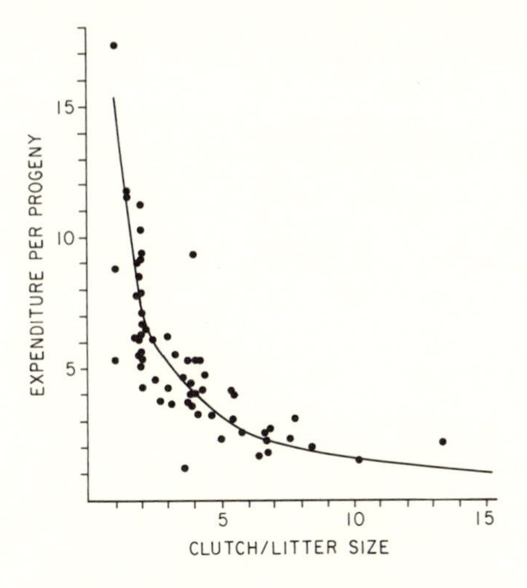


FIGURE 5.2. Average expenditure per progeny, expressed as a percentage of female body weight, is plotted against the average size of a clutch or litter for 62 species of desert lizards representing ten different families.

penditure per progeny is substantially improved (partial correlation coefficient = .704), an indication, once again, that these three aspects of reproductive tactics are by no means independent of one another. Indeed, pairwise partial correlation coefficients between the logarithms of these three variates are all nearly perfect ($r_{xy \cdot z}$ = .910, .924, and −.969). Species fall neatly on a plane in this three-space, as evidenced by a principal component analysis using log-transformed variables, which shows that the first two principal components reduce variance by a full 99.2%.

Frequency distributions of average clutch/litter sizes, total investment ratios, and expenditure per progeny among the species listed in Appendix E are shown in Figure 5.3. Note that expenditure per progeny varies over more than an order of magnitude, from 1% to 17% of a gravid female's body weight. Interestingly, species with narrow diets often, though certainly not always, tend to have higher than average expenditures per progeny. Two of the species with the highest expenditures per progeny, *Typhlosaurus gariepensis* and *T. lineatus*, probably experience intense competition: (1) these live-bearing, sub-

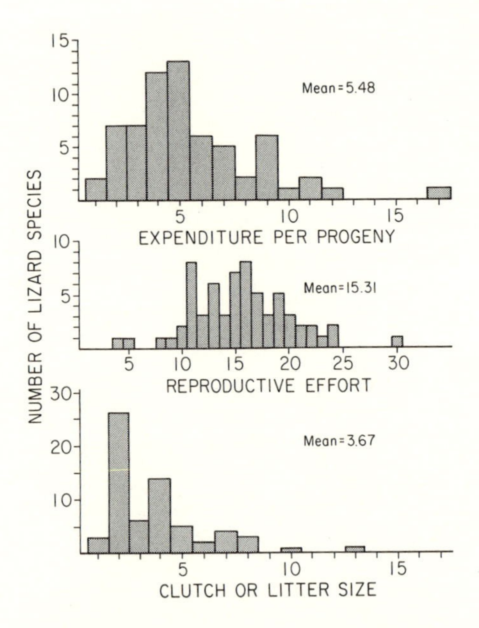


FIGURE 5.3. Histograms of numbers of species with different values for each of the three reproductive parameters discussed in the text.

terranean skinks exist at very high population densities; (2) individuals are long-lived with delayed maturity; (3) litter sizes are extremely small (means of 1.0 and 1.5, respectively); and (4) females very likely reproduce only biennially (Huey et al. 1974; Huey and Pianka 1974). These two Kalahari fossorial skinks are also both extreme food specialists, eating termites to the virtual exclusion of all other prey. The extremely high expenditure per progeny of *Typhlosaurus* may well be necessary to confer newborn animals with competitive ability sufficient to establish themselves in the highly competitive underground environment. Limited evidence indicates that investment per progeny is indeed responsive to and indicative of a lizard's competitive environment. Thus in *Typhlosaurus lineatus*, offspring are significantly heavier (and expenditure per progeny significantly greater) where this species occurs in sympatry with *T. gariepensis*, as compared with allopatric populations (Huey et al. 1974). Other food-specialized species seem also to encounter intense competition: among Australian geckos, species with relatively restricted termite diets tend to lay comparatively larger eggs and hence have higher expenditures per progeny than do those with more catholic diets (Pianka and Pianka 1976). A similar phenomenon appears to occur in the semiarboreal African skink *Mabuya spilogaster*: on one study area, it is syntopic with an ecologically very similar species, *Mabuya striata*. Expenditure per progeny in *M. spilogaster* increases significantly ($t$-test, $P < .01$) from allopatry (mean $= 4.39 \pm 0.21$, $N = 51$) to sympatry (mean $= 5.63 \pm 0.48$, $N = 19$).

Differences between viviparous and oviparous species are relatively slight, although, as noted above, viviparous species appear to invest slightly more in reproduction. Statistically significant differences exist between diurnal and nocturnal species of lizards in these reproductive statistics: nocturnal species have significantly smaller clutch/litter sizes and lower total investment in reproduction, but significantly higher expenditure per progeny (Table 5.2). These differences between diurnal and nocturnal species stem largely from a simple historical or taxonomic basis, since geckos and pygopodids dominate the nocturnal saurofauna and have a fixed clutch of only one or two eggs. Two viviparous nocturnal skink species (genus *Egernia*) also tend to have small litters; however, a third oviparous nocturnal skink, *Eremiascincus richardsoni*, does not have a small clutch size.

Reproductive tactics can be mapped, to a limited extent, onto the spatial-temporal thermoregulation axis plotted in Figure 3.8. Simple pairwise product-moment correlation coefficients between the three

TABLE 5.2. Comparisons of reproductive tactics of 24 nocturnal and 39 diurnal species, based on averages for each species.

| STATISTIC | NOCTURNAL SPECIES | | DIURNAL SPECIES |
|---|---|---|---|
| Clutch/litter size | | | |
| Mean | 2.03 | t = −5.55 | 4.83 |
| St. dev. | .72 | P < .001 | 2.40 |
| Reproductive effort | | | |
| Mean | 12.96 | t = −3.38 | 16.73 |
| St. dev. | 3.98 | P < .005 | 4.48 |
| Expenditure per progeny | | | |
| Mean | 6.70 | t = 4.43 | 4.22 |
| St. dev. | 2.15 | P < .001 | 2.16 |

NOTES: Nocturnal species constitute geckos, pygopodids, *Xantusia*, and several species of Australian skinks *(Egernia, Eremiascincus)*. The subterranean species of *Typhlosaurus* skinks are excluded.

reproductive variables and the slope/intercepts of body temperature regressions on air temperature are weak although generally statistically significant (Table 5.3). The strongest correlations seem to arise largely as a result of the small clutch sizes, steep slopes, and low body temperature intercepts of nocturnal lizards.

TABLE 5.3. Pearson product-moment correlation coefficients among slopes and intercepts of body temperature-air temperature regressions versus several reproductive tactic parameters (based on 63 species).

| REPRODUCTIVE TACTIC PARAMETER | CORRELATION WITH | |
|---|---|---|
| | *Slope* | *Intercept* |
| Clutch/litter size | −.32† | .46‡ |
| Reproductive effort | −.34† | .26* |
| Expenditure per progeny | .13ns | −.49‡ |
| Logarithm, clutch size | −.55‡ | .61‡ |
| Logarithm, expenditure per progeny | −.76‡ | −.39† |

NOTE: Significance levels are as follows:
ns = not significant
* = $P < .05$ † = $P < .01$ ‡ = $P < .001$

# 6 Natural History Miscellanea

Lizard tails have diversified greatly and serve a wide variety of functions for their possessors. Many climbing species, such as the Australian sandridge agamid *Gemmatophora longirostris*, have evolved extraordinarily long tails that serve as effective counterbalances. Tail removal experiments (Snyder 1952) have shown that such long tails also enable lizards to raise their forelegs up off the ground and to run on their hind legs alone (bipedality is a faster means of locomotion than tetrapodality). Prehensile tails are used as a fifth leg in climbing by other arboreal lizard species such as some geckos (e.g., *Diplodactylus elderi*) and by the true chameleons such as *Chameleo dilepis* of the Kalahari.

In several members of the Australian gekkonid genus *Diplodactylus* (subgenus *Strophurus*) (*D. ciliaris, D. elderi, D. strophurus*, and relatives), glandular tails secrete and store a smelly noxious mucous. When disturbed, these lizards squirt out large amounts of sticky odoriferous gorp (Rosenberg and Russell 1980). Surprisingly, tails of these geckos are fragile and easily shed (but quickly regenerated). A related Australian desert gecko *Diplodactylus conspicillatus* has a nonglandular but very short and stubby bony tail: these nocturnal termite specialists hide in the vertical shafts of abandoned spider holes during the day, and it is thought that they point head downwards, using their tails to block off these tunnels. Another Australian desert lizard with a similar yet different tail tactic is the climbing skink *Egernia depressa*. These lizards wedge themselves into tight crevices in mulga tree hollows (and rocks), blocking off the entrance with their strong and very spiny tails. Spinily armored tails are used by numerous other species of lizards in a similar fashion, including the Mexican iguanid *Enyaliosaurus clarki* and the Saharan agamid *Uromastix acanthinurus*.

Members of another bizarre group of Australian lizards (genus *Nephrurus*) possess a unique round knob at the tip of their tails. These large nocturnal lizards eat big prey, including other species of geckos on occasion. Both sexes carry the curious knob, but its function remains a total mystery. Unlike most geckos, their tails are *not* exceedingly fragile.

In some species of lizards (especially among juveniles), tails are brightly colored and/or very conspicuous, evidently functioning to lure a potential predator's attack away from the more vulnerable and less dispensable parts of the animal. Thus, when approached or followed by a large animal, the zebra-tailed lizard of the western North American deserts, *Callisaurus draconoides*, curls its tail up over its hindquarters and back, exposing the bold black-and-white pattern underneath, and coyly wriggles its tail from side to side. If pursued farther, zebra-tailed lizards resort to extreme speed (estimated at up to 20–30 km./hr.) and long zigzag runs. An Australian desert skink, *Ctenotus calurus*, lashes and quivers its bright azure-blue tail alongside its body continuously as it forages slowly through the open spaces between plants. Similarly, tiny *Morethia butleri* juveniles twitch their bright red tails as they move around in the litter beneath *Eucalyptus* trees.

Tails of many, but by no means all, lizards break off easily (indeed, some species can actually lose their tails voluntarily with minimal external force in a process known as autotomy). Freshly dismembered tails or pieces thereof typically thrash around wildly, presumably attracting a predator's attention while the former owner quietly slips away unnoticed (Dial and Fitzpatrick 1983). Certain small predators, such as the pygmy varanids *Varanus gilleni* and *V. caudolineatus*, may actually "harvest" the exceedingly fragile tails of geckos that are too large to subdue intact (Pianka 1969b). Some skinks (Clark 1971), including many *Ctenotus* (personal observation), return to the site where their tail was lost and swallow the remains of their own tail! Few, if any, other vertebrates display auto-amputation and self-cannibalism.

Many such lizards possess special adaptations for tail loss, including weak fracture planes within each tail vertebra, muscular attachments that facilitate autotomy and tail movement after dismemberment, as well as mechanisms for rapidly closing off blood vessels and healing. Losing its tail seems to have surprisingly little effect on a lizard, as individuals often resume basking and foraging within minutes as if nothing had happened. In such lizard species, of course, tails are quickly regenerated from the stub. Although regrown tails are occasionally almost indistinguishable externally from the original, their internal support structure is cartilaginous rather than bony.

Not all lizard tails are easily broken, however. Whereas most iguanids have fragile tails, their close relatives the agamids generally do not. Tails of varanids and true chameleons do not break easily, either. Lizards with such tough tails usually cannot regenerate a very complete tail if their original should happen to be lost. The evolutionary bases for these differences, sometimes between fairly closely related groups of lizards, are evasive.

The potential of tail-break frequency as an index to the intensity of predation on lizard populations was noted long ago by Haldane and Huxley (1927) and has since been used to attempt to estimate the amount of predation, although there are some problems and limitations with the procedure (Schoener 1979). Efficient predators that leave no surviving prey obviously will not produce broken tails but nevertheless may exert substantial predation pressures: broken and regenerated tails may therefore reflect lizard escape ability or predator inefficiency better than intensity of predation.

In western North America, predator densities increase from north to south (Table 6.1; Pianka 1967; Schall and Pianka 1980). Correlated with this latitudinal increase in predation, frequencies of broken and regenerated tails are higher at southern sites than at northern localities among four of the five widely distributed lizard species (Pianka 1967). In the well-studied species *Cnemidophorus tigris*, frequency of broken tails decreases with latitude (Figure 6.1); moreover, diversity of pred-

TABLE 6.1. Estimated densities (number per kilometer) of avian and snake predators on ten North American study areas, with approximate latitudes.

| Site | Latitude | Bird Predator Density | Snake Predator Density | Total Kilometers Walked |
|---|---|---|---|---|
| I | 42°12′ | .000 | .017 | 56 |
| L | 40°12′ | .044 | .033 | 92 |
| G | 38°48′ | .000 | .012 | 80 |
| V | 37°05′ | .000 | .057 | 122 |
| P | 36°18′ | .022 | .037 | 134 |
| S | 35°18′ | .071 | .058 | 103 |
| M | 35°06′ | .098 | .054 | 224 |
| T | 34°08′ | .120 | .069 | 251 |
| W | 33°41′ | .098 | .065 | 153 |
| C | 32°57′ | .098 | .070 | 328 |

ator escape behaviors utilized among members of these various *Cnemidophorus* populations also increases with the frequency of broken and regenerated tails (Figure 6.2; Schall and Pianka 1980). A greater variety of escape tactics, a form of behavioral "aspect diversity" (Rand 1967; Ricklefs and O'Rourke 1975), presumably reduces the ease with which predators can capture lizard prey.

Juvenile *Eremias lugubris* have evolved an interesting defense against predators involving Batesian mimicry of a noxious beetle (Huey and Pianka 1977a). These defenseless small Kalahari lacertid lizards mimic certain carabid beetles that emit pungent acids, aldehydes, and other repulsive chemicals when disturbed. (The local Afrikaans name for these beetles is "oogpister," which translates euphemistically as "eye squirter.") Adult *E. lugubris* are pale red and buff in color, matching the Kalahari sands, whereas juveniles have pitch-black bodies with white spots and a reddish tail. Adult lizards walk in a normal lizard gait, with their vertebral columns undulating from side to side, but juveniles walk stiff-legged, with their backs arched vertically and tails held flat against the ground (presumably this posture and the red color make their tails difficult to detect on the red Kalahari sands). If pur-

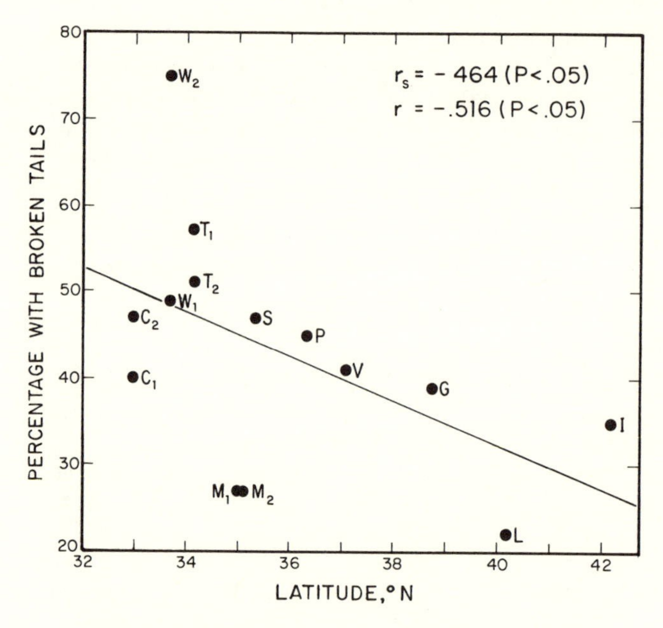


FIGURE 6.1. Percentage of *Cnemidophorus tigris* with broken and/or regenerated tails tends to decrease with increasing latitude.

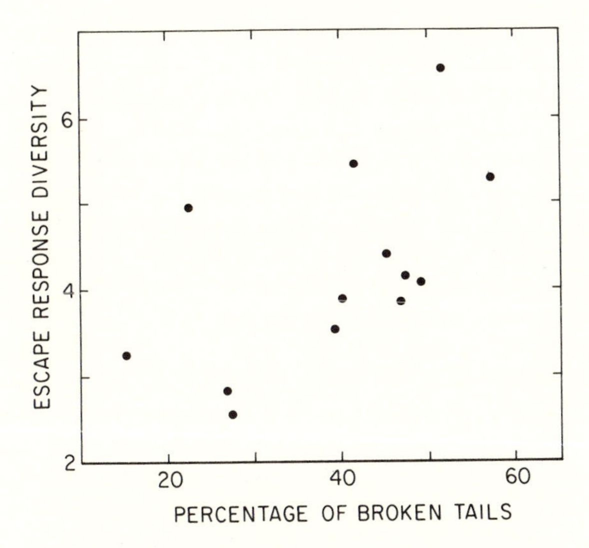


FIGURE 6.2. Diversity of escape responses tends to be greater in populations of *Cnemidophorus tigris* that exhibit higher frequencies of broken and regenerated tails.

sued, young *E. lugubris* abandon their "beetle walk" and dart rapidly for cover, using normal lizard locomotion. When they reach a snout-vent length of about 45 to 50 mm. (about the size of the largest oogpister beetles), these lizards "metamorphose" into the cryptic adult coloration and permanently abandon the arched walk. Interestingly, the incidence of broken and regenerated tails is lower in juvenile *E. lugubris* than it is among congeneric lacertids in the same habitats, suggesting that this beetle mimicry may successfully reduce predatory attacks.

A very important feature of the Australian sandy deserts is the complex burrows of *Egernia striata* (Pianka and Giles 1982); these nocturnal skinks dig elaborate tunnel systems that are used as diurnal retreats by many species of nocturnal geckos, including *Heteronotia binoei*, *Nephrurus levis*, and *Rhynchoedura ornata*. *Egernia* excavations are also exploited as refuges from predators and the elements by various diurnal lizards, including *Ctenophorus isolepis* and *Varanus eremius* (see also below). Large elapid snakes (*Pseudechis australis* and *Pseudonaja nuchalis*) are also regularly encountered in these tunnel systems. *E. striata* burrows are elaborate, with several interconnected openings often as far as a meter apart and vaguely reminiscent of a tiny rabbit warren. Most of the sand removed from a *striata* burrow is piled up in a large mound outside one "main" entrance, which usually points south or southwest (Pianka and Giles 1982). Activity of *E. striata* is strongly seasonal, with 83% of 195 skinks being collected during the summer months of November through February.

Another sympatric but smaller and more crepuscular *Egernia* species, *E. inornata*, digs a much simpler burrow, consisting of a U-shaped tube with but one arm of the "U" open (this is the sole entrance and the only open exit to the burrow); the other arm of the "U" typically stops just below the surface of the ground and is used as an escape hatch by breaking through in an emergency. *E. inornata* individuals may often have two such burrows 10–20 meters apart. Burrows of *inornata* are usually about a third of a meter beneath the surface at their deepest spot, and are somewhat shallower than those of *striata*. The sand removed from *inornata* burrows is typically spread out in a thin, fan-like layer radiating away from the entrance (lizards have been observed pushing sand out and smoothing it over with their forefeet and then backing down into their hole). In contrast to *striata* burrows, this entrance most often faces north or northwest (Pianka and Giles 1982). In addition to being more diurnal than *E. striata, Egernia inornata* are not as strongly seasonal in activity (only 44% of 131 skinks were collected in the four months from November through February). Thus *E. inornata* is appreciably more active during colder weather.

Why do the two species construct such different burrows and why do they have such curious compass orientations? Moving all of the sand out of a single opening of an extensive and complex *striata* burrow system would seem to entail considerable extra energetic expenditure: what could be the counterbalancing benefits? Deep sand is a darker red than surface sand and such tailings give away the positions of burrow entrances. Perhaps consolidating all such conspicuous diggings in one massive pile reduces the likelihood that they will attract undesirable attention (these mounds themselves are sometimes hidden inside *Triodia* tussocks). Alternatively, the mound itself could serve as a convenient lookout and/or basking platform (the southerly orientation might facilitate the latter function by providing a sloping surface roughly perpendicular to the sun's rays from the north). *E. striata* individuals are occasionally encountered basking during daylight hours. The north-facing entrances to *inornata* tunnels are more difficult to explain. One possibility is that a lizard sitting in such an entrance during the day would be exposed to the relatively warm northern sky. Thus interpreted, the interspecific difference between these two *Egernia* in orientation of their burrows would reflect the observed difference in seasonality and extent of diurnal activity.

Australian *Varanus* are exceedingly wary, essentially unapproachable and unobservable lizards, making them a real challenge to study. Fortunately, they leave fairly conspicuous tracks, and one can deduce quite a lot about their natural history from careful study of this spoor. I must admit that these magnificent and exceedingly intelligent lizards are my favorites. The largest species, *V. giganteus*, reaches two meters or more in total length, whereas some of the smaller "pygmy goannas," such as the ubiquitous and very important lizard predator *V. eremius* (Pianka 1968), achieve lengths of only about 40 cm. Two other species, *V. gouldi* and *V. tristis*, are intermediate in size (Pianka 1970b, 1971b). All four of these species range over extensive areas and consume very large prey items, particularly other vertebrates (especially lizards). *V. tristis* and two other little-known small species, *V. caudolineatus* and *V. gilleni*, are arboreal, whereas the other three species (*V. eremius, V. gouldi*, and *V. giganteus*) are terrestrial. Each species leaves its own very distinctive track. Daily forays typically cover a distance of one kilometer.

*V. gouldi* appears to encounter most of its food—predominantly lizards and reptile eggs—by digging; it seems to have very keen powers of olfaction. Stomachs of 63 of these large monitors contained 21 other species of smaller lizards as prey (35% of the overall diet by volume).

*V. tristis* also consumes other lizards as well as baby birds (and probably bird eggs); its very distinctive track typically runs more or less directly from tree to tree (presumably the monitors climb most of these trees looking for food). I was fortunate to observe a direct confrontation between a breeding Galah Cockatoo parrot (*Cacatua roseicapilla*) and one of these varanid lizards. I first heard a Galah screeching loudly somewhere nearby, but not at me. When I saw the bird, it was screaming loudly as if in great distress, with its crest held high and wings outstretched. On the ground when first sighted, the Galah flew up onto a fallen log under a *Eucalyptus* tree and then into the tree, which later proved to be its nesting tree. On my approach, a large adult *V. tristis* clambered over the same log toward the tree. While the parrot continued to screech and began harassing the lizard, I approached. In an instant, the lizard was up the tree and out of sight around the other side. At this point, the Galah actually attacked the goanna (about 3 meters off the ground), and drove it back down. The incident suggests that these large climbing lizards must be a potent threat to hole-nesting parrots. *V. tristis* activity is highly seasonal and the animals seem to rely on building up fat reserves during times of plenty to get them through lean periods. *V. tristis* rely primarily on other species of lizards for food (71% of the diet by volume).

Judging from the frequency of its unique and conspicuous tracks, *V. eremius* is common in Australian sandy deserts. Unlike the larger goannas, it is active all year long. This beautiful little red *Varanus* is extremely wary and hence seldom seen, however. Nevertheless, a great deal about its activities can be inferred from its tracks: the comments to follow are based upon impressions gained while following hundreds of kilometers of *eremius* tracks. Individuals usually cover great distances when foraging and I have often followed a fresh track for distances of up to one kilometer. Tracks indicate little tendency to stay within a delimited area; home ranges of these lizards must be extremely large. These small goannas are attracted to fresh holes and diggings of any sort, and almost invariably will visit any man-made digging within a day or two after it is made. In contrast to *V. gouldi*, *V. eremius* seldom do any digging for their prey, but rather rely upon catching it above ground. I have more than once noted an *eremius* track intercept the track of another smaller lizard with evidence of an ensuing tussle. Once, an *eremius* was actually observed to attack another lizard from ambush: on this occasion, a large *eremius* leapt out of a loose *Triodia* tussock when a small blue-tailed skink (*Ctenotus calurus*) came within a few centimeters of the edge of the tussock. Stomach contents reveal that other lizards comprise over 70% of the *eremius* diet by volume, whereas large grasshoppers plus an occasional large cockroach or scorpion constitute most of the remainder. Nearly any other lizard species small enough to be subdued is eaten (60 stomachs with food contained 42 individual lizard prey representing some 14 other species). In a typical foraging run, an individual *eremius* often visits and goes down into several burrows belonging to other lizard species (especially the complex burrow systems of *Egernia striata* described above). These activities could be in search of prey, related to thermoregulatory activities, and/or simply involved with escape responses. Certainly an *eremius* remembers the exact positions of the burrows it has visited, since it almost inevitably runs directly to the nearest one when faced with an emergency.

# 7 Analysis of Community Structure: Theory and Methods

A major challenge facing ecologists is to understand just *how* natural communities are in fact put together. Only then will we be able to begin to ask meaningful questions about *why* they have particular observed properties. Perhaps the most vital unresolved problem is whether or not collections of species possess truly emergent properties that transcend those of mere collections of populations: for example, are patterns of resource utilization among component species' populations "coadjusted" so that they mesh together in a meaningful way? Techniques for simply describing many potentially important attributes of community structure remain in their infancy, however. Operational ways of measuring numerous community-level phenomena have not yet been fully worked out. As Orians (1980) points out, choice of appropriate "macro" descriptors or "aggregate variables" is essential to progress at the community level of organization: community ecology both depends upon and yet is simultaneously constrained by the identification of such conceptual building blocks. To be most useful, these macrodescriptors must oversimplify population-level processes while retaining their essence. In this chapter, some such aggregate variables are identified, methods of analysis are outlined, and some theory is briefly discussed. The following two chapters apply these methods to my lizard data base. Perhaps they will underscore some of the above assertions. I hope also to extend and refine methods of community analysis somewhat, as well as contribute basic information on the structure of these particular natural assemblages.

During the last half century the term "niche" has been used in a wide variety of different ways—so many, in fact, that some population biologists prefer not to use the word at all. However, the niche concept has gradually become linked with species-specific resource utilization and with the notion that species avoid interspecific competition. A growing school of American population biologists now equates the ecological niche with patterns of resource use. This chapter and the next two center around this definition, although I personally subscribe to a somewhat broader view of the niche.

A fairly extensive theoretical literature has been developed in which consumers are modeled with bell-shaped resource utilization curves. Consumer species within "competitive" communities are visualized as being at equilibrium with their resources, with the total rate of consumption of renewable resources being exactly balanced by their rates of renewal (Figure 7.1). An analogous view of nonrenewable resources (such as space) is that available resources are used to the fullest extent possible, or that the habitat is saturated with consumers (in this case the vertical axis of Figure 7.1 would be scaled in amount of resource used rather than in renewal rates). Discrete resources that cannot be ordered on a linear axis, such as prey taxa, are easily treated similarly.

Ecologists are divided in their attitudes concerning the impact of competition on community structure (Wiens 1977; Connell 1980; Schoener 1982, 1983b; Strong et al. 1984; Price et al. 1984). While the debate rages, many believe that the importance of predation has been underestimated due to overenthusiastic acceptance of the com-

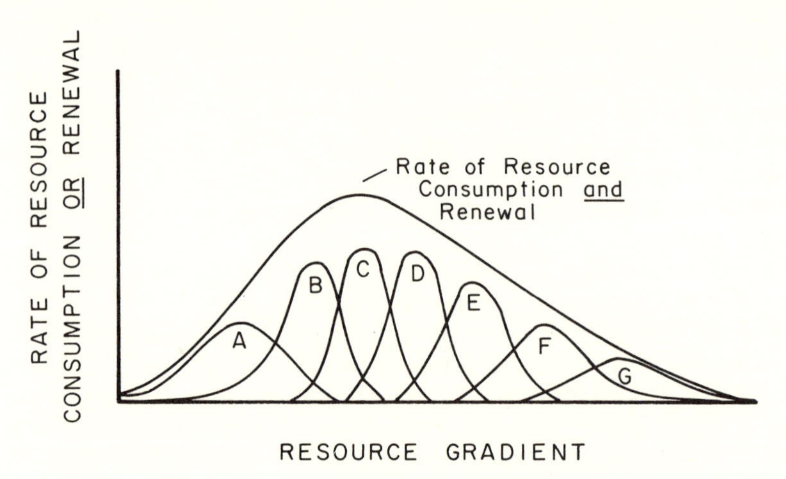


FIGURE 7.1. A "competitive community" is often visualized as one in which consumer species are at equilibrium with one another and with their resources, so that the total rate of consumption of resources is balanced by their renewal rate across the entire spectrum of available resources.

petitive community dogma. Some insist that only direct experimental manipulations can be brought to bear on this dilemma. Others have pointed out that even experiments have serious limitations and may easily be misinterpreted (see, for example, Bender et al. 1984). Resolution of this long standing dispute over the strength of competition in structuring natural communities will not be easy. In any case, the competitive view of community structure provides a conceptual framework that is appealingly simple and operationally tractable, which leads quite naturally to several extremely useful notions including niche breadth and niche overlap.

Substantial decisions must be made at the outset of virtually every ecological study. Which resource categories are most appropriate? How many different resource states should a researcher recognize? (The obvious immediate answer "As many as possible" proves to be overly facile and erroneous.) Of course, if resource categories are too broad, patterns of differential use will be obscured and ecological similarity will be overestimated (Greene and Jaksic' 1983). On the other hand, recognition of extremely narrow resource categories can result in problems of the opposite sort (i.e., meaningless differences may be generated).

Problems of selecting resource categories can be illustrated by *Varanus*. In my samples, *Varanus eremius* consumes 14 other species of lizards, whereas *V. gouldi* eats 21 other species. *V. tristis* is recorded as eating 11 other lizard species. All three of these monitors are lizard eaters, taking a total of 31 other species of lizards as prey. However, 19 of these 31 prey species are eaten by only one *Varanus* species. In my samples, only 3 other species were actually eaten by all three monitors (*eremius* and *gouldi* share 6 prey species; *eremius* and *tristis* have 4 prey species in common; and *gouldi* and *tristis* share 8 prey). I strongly suspect that, given an opportunity, any of these *Varanus* would readily catch and consume almost *any* other species of lizard that could be subdued. Ideally, my sample sizes of *Varanus* would be in the thousands rather than sixties, and observed dietary differences would be genuine. Clearly, in this imperfect case, the most appropriate prey category to recognize is simply "other lizards." Thus one must rely on prior knowledge of the organisms concerned as well as on biological intuition in choosing resource categories to recognize.

If resources can be ordered (such as prey size, height above ground, etc.), niche breadth is readily quantified by any of various measures of dispersion, like the variance or range of the utilization curves. Even if resources cannot be arranged along such a sequence but nevertheless occur in discrete categories (prey species or categories of items like termites, beetles, and ants), various measures of dispersion are still possible, such as Shannon's (1949) information theoretic index of uncertainty:

$$H = -\sum_{i=1}^{n} p_i \ln p_i, \qquad (1)$$

where $i$ subscripts resource categories, $p_i$ is the proportion of resource in the $i$ class, and $n$ is the total number of categories. If all $i$ resource classes are represented equally (all $p_i = 1/n$), this measure reaches a maximal possible value of $\ln(n)$, designated $H_{max}$. A standardized $H/H_{max}$ (sometimes called $J$) thus has a range of from near zero to one. Another very useful measure of the dispersion of nonorderable categories, used here to quantify diversity and niche breadth, is the diversity index first proposed by Simpson (1949):

$$D \text{ (or } B) = \frac{1}{\sum_{i=1}^{n} p_i}, \qquad (2)$$

where symbols are as before. This measure varies from 1 to $n$, and when divided by the number of species or resource categories, $n$, also has a range of from near zero (actually $1/n$) to unity. In practice, such measures are usually computed from utilization data summed over a sample of individuals: as such, they represent populational niche breadths.

Two components of niche breadth can be distinguished (Van Valen 1965; Roughgarden 1972, 1974). The within-phenotype component refers to the variety of resources used by an average individual, whereas the between-phenotype component is based on differences between individuals in resource utilization, increasing as individual consumers diverge from one another in their use of resources. The two components sum to give the total populational niche breadth. A population with a niche breadth determined entirely by between-phenotype variation would be composed of specialized individuals with little niche overlap among them. A population composed of pure generalists with each member exploiting the full range of resources used by the entire population would have a between-phenotype component of niche breadth of zero and a maximal within-phenotype component. Real populations presumably lie somewhere between these two endpoints but have seldom been quantified in these terms. Numerous factors influence niche breadth, including densities of available resources as well as their spatial and temporal distribution. Resource specialization (and hence narrow niche breadth) is favored by high resource density as well as by patchily distributed resources that are clumped in locally concentrated areas.

In the absence of any competitors or other enemies, the entire set of resources used is referred to as the "fundamental" niche (Hutchinson 1957) or the "virtual" niche (Colwell and Futuyma 1971). Any real organism seldom, if ever, exploits its entire fundamental niche, since its activities are almost always somewhat curtailed by its competitors as well as by its predators; hence, its "realized" or "actual" niche is a subset of the fundamental niche. Considerations of the variety of factors influencing niche breadth lead into the problem of specialization versus generalization (Levins 1968; Joern 1979; Fox and Morrow 1981; Greene 1982). A fairly substantial body of theory on optimal foraging predicts that dietary niche breadth should generally increase as food availability decreases (Emlen 1966, 1968; MacArthur and Pianka 1966; Schoener 1971; MacArthur 1972; Charnov 1976). One observation in support of such arguments was reported earlier: recall that the diversity of foods eaten by the North American teiid *Cnemidophorus tigris* varies inversely with recent rainfall (hence presumably food supplies); see Figure 4.4 (page 54).

A competitor can act either to compress or to expand the realized niche of another species, depending upon whether or not it reduces resource levels uniformly (which would lead to niche expansion) or in a patchy manner (which should often result in a niche contraction). Literature on the effects of intraspecific versus interspecific competition on niche breadth is extensive (for an entry, see MacArthur 1972 and/or Schoener 1974a).

The basic raw data for analysis of niche overlap is the resource matrix, which is simply an $m$ by $n$ matrix indicating the amount (or better yet, the rate of consumption) of each of $m$ discrete resource states utilized by each of $n$ different consumer species (Table 7.1). Utilization coefficients ($u_{ij}$) in the resource matrix are often converted into proportions or "$p_{ij}$," which sum to unity for each consumer (here the $i$'s subscript resource states from 1 to $m$ and the $j$'s subscript species from 1 to $n$). Such resource matrices for desert lizards, exemplified in Appendices C and E for microhabitat and diet, are rich in structure, with numerous zero elements. Perhaps more interesting are the very large entries for some species of consumers on certain resource states. These are considered in Chapters 3 and 4 and are discussed further in Chapters 8 and 9.

An alternative procedure, perhaps far more preferable but also much

TABLE 7.1. The form of the resource matrix.

| RESOURCE STATE | CONSUMER SPECIES 1 | 2 | 3 | 4 | • | • | • | *n* |
|---|---|---|---|---|---|---|---|---|
| 1 | $u_{11}$ | $u_{12}$ | $u_{13}$ | $u_{14}$ | • | • | • | $u_{1n}$ |
| 2 | $u_{21}$ | $u_{22}$ | $u_{23}$ | $u_{24}$ | • | • | • | $u_{2n}$ |
| 3 | $u_{31}$ | $u_{32}$ | $u_{33}$ | $u_{34}$ | • | • | • | $u_{3n}$ |
| • | • | • | • | • | • | • | • | • |
| • | • | • | • | • | • | • | • | • |
| m | $u_{m1}$ | $u_{m2}$ | $u_{m3}$ | $u_{m4}$ | • | • | • | $u_{mn}$ |

NOTE: Utilization coefficients, $u_{ij}$, represent the rate of consumption of resource state $i$ by consumer species $j$.

more difficult in practice, is to use "electivities" (Ivlev 1961; Schoener 1974b; Lawlor 1980a), which reflect the degree to which each consumer's actual pattern of utilization deviates from that of a totally nonselective consumer (i.e., one that uses all resources directly in proportion to their actual availability). In effect, abundant resources and/or those that are used by many different consumers are given less weight than rare resources or those that are used by only a few of the community's consumers. Electivities are exceedingly difficult to estimate because the effective availability of any given resource state usually differs among consumers: as one example, nectar in flowers of a plant species with long corollas may be much more available to a hummingbird with a long beak than to another sympatric hummingbird species with a shorter beak, even though both bird species encounter the flowers equally often. (Of course, beak morphology evolves in response to patterns of resource availability.) The problem of estimation of availabilities is much more acute for some types of resources than it is for others.

In some circumstances, one can assume that community-wide utilization patterns approximate availabilities and simply use the row sums of the resource matrix as estimates of availability (e.g., see Lawlor 1980a, b). In such an approach (used below), common and/or large species exert a disproportionate influence, making it risky in terms of built-in circularity, but at the same time these "bioassays" of availabilities may be vastly superior to attempts to estimate availabilities in any other more direct way. Schoener (1974b) and Lawlor (1980a) make a case for examination of electivity-based measures of ecological similarity in preference to those that rely only upon proportional utilization, noting that the latter may be misleading because they are biased by relative resource availabilities. Clearly, patterns of resource utilization cannot be adequately understood without consideration of resource availabilities.

In any case, however, once obtained, such a resource matrix can be used to generate an $n$ by $n$ matrix of ecological overlap or "similarity" (Lawlor 1980a) with ones on the diagonal and values less than unity as off-diagonal elements, reflecting the degree of resource sharing between each pair of consumer species. Niche overlap has been quantified with a wide variety of different formulae (Horn 1966; MacArthur and Levins 1967; Levins 1968; Schoener 1968; Colwell and Futuyma 1971; MacArthur 1972; Pielou 1972; Vandermeer 1972; Pianka 1969b, 1973; Hurlbert 1978; Linton et al. 1981). Here overlaps are computed with a symmetric version of an equation first proposed by MacArthur and Levins (1967):

$$O_{jk} = \frac{\sum_{i=1}^{n} p_{ij} p_{ik}}{\sqrt{\sum_{i=1}^{n} p_{ij}^2 \sum_{i=1}^{n} p_{ik}^2}}, \tag{3}$$

where the symbols are as before, with $j$ and $k$ subscripting the species.[1] The same formula can also be used to compute "similarities" when $p_{ij}$ are replaced by estimates of "electivities" or $a_{ij}$ for various resource states rather than coefficients of proportional utilization among various consumer species (see Lawlor 1980a for discussion of the merits and techniques of this approach). Most overlap indices generate values that vary from zero to one so that at these endpoints of no overlap and total overlap, various measures are equivalent. Horn (1966) found departures from linearity among different overlap indices at intermediate amounts of overlap of only about 10%; hence the precise index used is presumably often somewhat arbitrary, barring mathematical arguments for the use of any particular form (see Abrams 1980; Colwell and Futuyma 1971; Pielou 1972; May 1974, 1975; Hurlbert 1978; and Linton et al. 1981 for discussion).

Overlap has sometimes been equated with competition coefficients or "alphas" because overlap is so much easier to measure. However, the caveat must be issued that overlap need not result in competition unless resources are in short supply. Extensive overlap may well be possible when a surplus of resources exists (such as during termite swarms after summer rains), whereas maximal tolerable overlap may

[1] For further evaluation of this and other overlap and similarity measures, see Horn (1966), May (1975), Hanski (1978), Hurlbert (1978), Abrams (1980), Lawlor (1980a), and/or Linton et al. (1981).

be much less in more saturated environments. The ratio of demand over supply should be fairly constant along any *particular* resource gradient because otherwise a local competitive vacuum would occur. Along any given resource spectrum, intensity of competition should therefore be directly proportional to actual overlap observed along that particular resource spectrum. But considerable caution must be exercised in comparing patterns of niche overlap along different resource axes or between different communities.

Avoidance of interspecific competition is but one of several mechanisms that may lead to resource partitioning; others include predator avoidance as well as innate design constraints experienced by consumers (Joern 1979; Toft 1985).

Niche overlap theory is usually framed in terms of a single niche dimension. As such, each species has only two immediate neighbors in niche space, as in Figure 7.1. Moreover, overlap matrices contain many zeros and only two large positive entries on the off-diagonal per row. Real plants and animals, however, seldom differ in their use of just one resource: rather, pairs of species usually overlap along two or more niche dimensions. As the effective number of niche dimensions rises, the number of potential neighbors in niche space increases more or less geometrically. Also, off-diagonal overlap elements of zero are fewer and the variance in observed overlap usually falls, both within rows and over the entire overlap matrix. Niche dimensionality also strongly affects the potential for "diffuse" competition arising from the total competitive effects of all interspecific competitors (MacArthur 1972). The overall effect of relatively low competitive inhibition per species summed over many other species could be as strong or even stronger than much more intense competitive inhibition (per species) by fewer competing species. An increased number of niche dimensions, by generating a greater potential for immediate neighbors in niche space, can thus intensify diffuse competition. Various aspects of niche dimensionality are considered by Yoshiyama and Roughgarden (1977) and by Rappoldt and Hogeweg (1980).

Imagine that height above ground and prey size are the two critical niche dimensions that species use differentially and thus avoid or reduce interspecific competition (Figure 7.2). Analysis of resource utilization and niche separation along more than a single niche dimension should ideally proceed through estimation of proportional simultaneous utilization of all resources along each separate niche dimension. For the case at hand, a three-dimensional resource matrix is necessary, with each entry representing the probability of capture of a prey item of a given size category at a particular height interval above ground

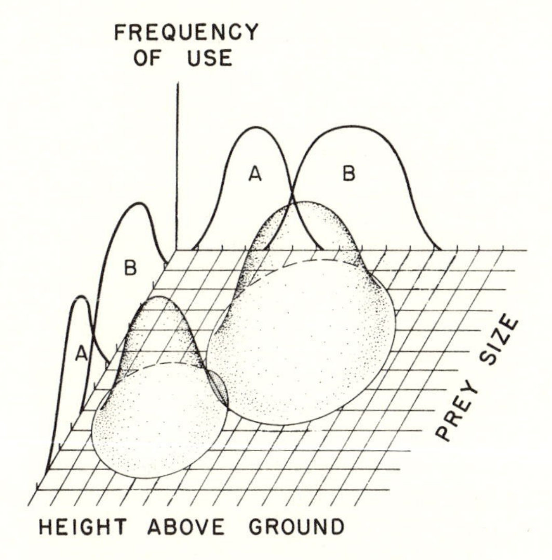


FIGURE 7.2. Diagrammatic representation of resource utilization by two hypothetical consumer species among two niche dimensions, prey size and foraging height. Although the shadows of the three dimensional peaks on each separate niche dimension overlap, true overlap in both dimensions is very slight.

by each of the consumer species present. Obtaining such multidimensional utilization data is extremely difficult in practice, however, because most animals move and integrate over both space and time. Accurate estimates of an animal's true use of a multidimensional resource space can be obtained only by monitoring continually an individual's use of all resources (even then, the degree to which prey individuals move between microhabitats will affect competition in obscure but vitally important ways). Since such continual observation is usually extremely tedious or even impossible, one usually must attempt to approximate from separate unidimensional utilization patterns (Figures 7.2 and 7.3). Just as the three-dimensional shape of a mountain cannot be accurately determined from two of its silhouettes as viewed at right angles, these unidimensional "shadows" do not allow completely accurate inference of the true multidimensional utilization (May 1975). The question of the degree of dependence or independence of dimensions assumes critical importance. Provided that niche dimensions are truly independent, with prey of any size being equally likely to be captured at any height, overall multidimen-

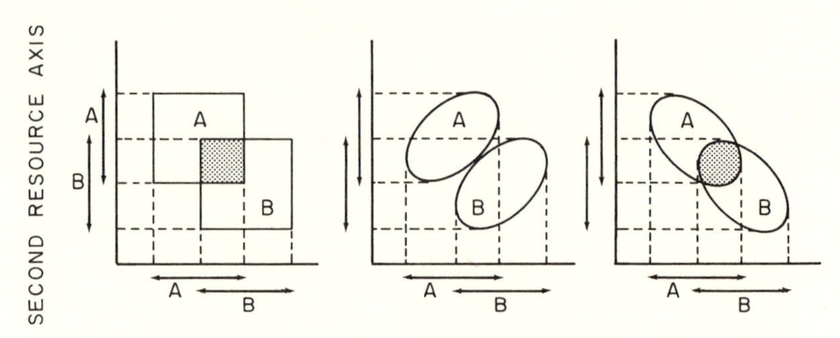


FIGURE 7.3. Three possible cases for the use of two resource dimensions, assumed to be constant within boxes or ellipses, for two hypothetical consumer species. Although unidimensional projections are identical in all three cases, true multidimensional overlap is zero in the central case. Truly independent resource dimensions are shown on the left, with any point along a given resource axis being equally likely along the entire length of the other resource axis; under such circumstances, niche dimensions are orthogonal, and unidimensional projections accurately reflect multidimensional conditions. However, when niche dimensions are partially dependent on one another (as in the center and right panels), unidimensional projections can be misleading.

sional utilization is simply the product of the separate unidimensional utilization probabilities (May 1975). Under perfect independence, the probability of capture of item $i$ in microhabitat $j$ is then equal to the product of the probability of being in microhabitat $j$ times the probability of capturing item $i$. Unidimensional estimates of various niche parameters (including breadth and overlap) along component niche dimensions may then simply be multiplied to obtain multidimensional estimates. However, should niche dimensions not be orthogonal but partially interdependent (Figure 7.3), there is unfortunately no substitute for knowledge of the true multidimensional utilization. True multidimensional overlap can vary greatly depending upon the precise form of this dependence (see Figure 7.3). In the extreme case of complete dependence (with, for example, prey of each size being found only at one height)—appearances to the contrary—there is in fact only a single niche dimension, and a simple arithmetic average provides the best estimate of true utilization. In fact, the mean of estimates of unidimensional overlap obtained from two or more separate unidimensional patterns of resource use constitutes an *upper bound* on the true multidimensional overlap (May 1975).

Because niches are usually multidimensional, it is exceedingly difficult to assimilate all the information contained in a matrix of overlap values for a community with even a moderate number of species. As A. Sheldon (pers. comm.) has noted, we badly need to find ways of representing community structure with simple pictures. Consider, for example, representing species as dots (or more realistically as clouds of dots or probability density distributions) in resource space (Pielou 1969). Two species can be plotted as points on a line with the distance between them—their niche separation—being inversely proportional to their niche overlap. Various measures of niche separation are possible; one is the Euclidean distance between species in as many dimensions as appropriate:

$$d_{jk} = \sqrt{\sum_{i=1}^{n} (p_{ij} - p_{ik})^2}, \quad (4)$$

where, as before, $p_{ij}$ and $p_{ik}$ represent proportions or electivities of resource state $i$ utilized by species $j$ and $k$, respectively. Another possible measure of niche separation is simply one minus overlap or similarity.[2] Three species can be represented similarly as the points of a triangle in two dimensions and four species as the vertices of a tetrahedron in three dimensions. But if several niche dimensions separate species, even a community of only five species may require a four-dimensional space to depict accurately all the interactions among its component species. A community of a dozen species separated along multiple niche dimensions clearly defies having its portrait painted! If dimensions are correlated, dimensionality can be reduced by changing coordinate systems with multivariate procedures such as principal components (Inger and Colwell 1977), discriminant functions (Green 1971, 1974; Dueser and Shugart 1979), factor analysis, or multidimensional planes (Aspey and Blankenship 1977; Holmes et al. 1979). Changing coordinate systems does not alter Euclidean distances, which still accurately reflect spacing patterns among species.

Another innovative and intriguing way of depicting some aspects of community structure uses graph theory (Sugihara 1982, 1984).

Some of the information contained in an overlap or similarity matrix can be conveyed using simple frequency distributions of the magnitudes of overlap values. Such histograms, however, do not adequately show the extent to which species are dispersed evenly throughout

[2] Euclidean distances between species are very strongly inversely correlated with overlaps (calculated with equation [3]). For a wide variety of data matrices, correlation coefficients average about −.90, ranging from −.82 to −.97. Effectively, overlap is the inverse of distance.

resource space versus clustered into groups of functionally similar species, such as foliage-gleaning insectivorous birds (termed "guilds" by Root 1967). Members of such guilds would interact strongly with one another but only weakly with the remainder of their community. Guilds are of interest because they presumably represent arenas with the most potential for intense interspecific competition. A useful technique that depicts some of the community's "connectedness" involves nearest-neighbor analysis: each species' neighbors in niche space are ranked from the nearest niche neighbor to the most distant (Inger and Colwell 1977). In such plots, very similar species fall out together, whereas species on the periphery of niche space with low overlap with the remainder of the community tend to fall well below other species. Mean overlap among all members of a community decays monotonically with neighbor's rank in niche space, of course. The standard deviation in overlap may, however, increase, decrease, or even rise and then fall as one moves from the nearest niche neighbor to the most distant. Inger and Colwell argue that humps in such standard deviation versus nearness rank curves are indicative of clustering and hence guild structure as follows: Because close-in neighbors in niche space tend to belong to the same guilds, standard deviation in overlap is primarily within guilds at low ranks and thus low. But if a community contains two or more guilds of different size, at a rank of one beyond that of the smallest cluster, standard deviation in overlap should increase because both within-guild overlap values and some between-guild values occur at the same rank. (At still higher ranks, standard deviation in overlap falls because all overlaps are now between members of different clusters.) Thus such humps in standard deviation curves are partially attributable to the presence of two or more distinct clusters of species (guilds).

Using the "single-linkage" criterion of cluster analysis allows objective definition of a guild as a group of species, each of which is separated from all other such clusters by a distance greater than the greatest distance between the two most disparate members of the guild concerned. This conservative definition allows complex hierarchical patterns of nesting of smaller guilds within larger ones. Less conservative "multiple-linkage" algorithms have also been developed for cluster analysis. Unfortunately, all such clustering techniques necessarily distort spacing patterns since multiple dimensions must be collapsed to a single dimension.

Ecologists have long been intrigued with the notion that there should be an upper limit to how similar the ecologies of two species can be and still allow coexistence. Among the many concepts that have emerged from such thinking are the "principle" of competitive exclusion, character displacement, limiting similarity, species packing, and maximal tolerable niche overlap (for some examples, see Hutchinson 1959; Schoener 1965; MacArthur and Levins 1967; MacArthur 1970; May and MacArthur 1972; Pianka 1972; Grant 1972; Abrams 1975, 1976, 1977). A good number of models of niche overlap in competitive communities have been developed (MacArthur and Levins 1967; MacArthur 1970; May and MacArthur 1972; May 1974; Gilpin 1974; Rappoldt and Hogeweg 1980; Roughgarden 1974, 1976; Yoshiyama and Roughgarden 1977). Inger and Colwell (1977) argue that a low variance in overlap for very close neighbors in niche space supports the idea of limiting similarity; however, this pattern might also arise as an artifact of the necessary upper bound of unity on overlap values (at low nearness ranks, neighbors are closer to this bound than they are at more distant ranks).

Saturated communities can differ in diversity in only three ways, which are not mutually exclusive (MacArthur 1965, 1972): (1) The diversity of available resources determines the variety of opportunities for ecological diversification within a community. Communities with fewer types of resources will not support as many species as those with a greater variety of resources, all else being equal. (This represents a "smaller overall niche space" or "fewer niches.") (2) As the diversity of utilization of resources by an average species increases, the number of species that can coexist within a community must decrease. (This corresponds to "larger niches.") (3) Two communities similar in both of the above respects can still differ in diversity, if they differ in the average extent to which resources are shared, or in the amount of niche overlap. A community with greater overlap will support more species than one with less overlap simply because more species use each resource. (This represents "smaller exclusive niches.") Hence diversity should decrease with increases in the niche breadths of a community's component species, but diversity should increase with the range of available resources and with the extent of tolerable overlap.

MacArthur (1972) derived the following simple equation to describe the influence of these various factors on community diversity:

$$D_S = \frac{D_R}{D_U}(1 + C\bar{\alpha}), \tag{5}$$

where $D_S$ represents species diversity, $D_R$ is the overall diversity of resources used by the entire community, $D_U$ represents the average diversity of utilization (niche breadth) of each species, $C$ measures the number of neighbors in niche space, and $\alpha$ is the mean "competition coefficient" (I prefer to consider this the average degree of resource sharing or niche overlap). While essentially tautological in its derivation, MacArthur's elegant community equation does underscore the importance and modes of action of resource diversity, niche breadth, niche dimensionality, and niche overlap in considerations of species diversity.

### *Null Hypotheses*

Until fairly recently, students of resource partitioning and community structure were unable to accomplish much more than simply describe existing patterns of resource partitioning among various coexisting consumer species. Even such descriptive efforts seldom allow very useful comparisons with other studies of communities, partially because there is "no . . . standard protocol for community analysis" (Inger and Colwell 1977). In some cases with low niche dimensionality, observed estimates of overlap have been compared with values of limiting similarity generated from various theoretical arguments such as those of MacArthur and Levins (1967) or May and MacArthur (1972). (For examples, see Orians and Horn 1969 and/or May 1974.) But such comparisons have not been particularly revealing since values of limiting similarity depend strongly on the specific assumptions of models concerned (Abrams 1975).

Sale (1974) responded to this dire need for null hypotheses by suggesting that communities might be compared to randomized versions of themselves. Overlap in observed communities of grasshoppers did not differ markedly from that in such randomized analogues, leading Sale to conclude that competition had not been a force in reducing overlap among these communities of insects or otherwise organizing them. In a similar analysis using data on some of my desert lizard communities, however, Lawlor (1980b) found that average similarity was substantially lower in observed communities than in randomized replicates, suggesting that competition has shaped the organization of these lizard communities. This rather promising "neutral model" approach has now been exploited in a goodly number of studies of community structure (Caswell 1976; Colwell and Winkler 1984; Inger and Colwell 1977; Pianka, Huey and Lawlor 1979; Strong, Szyska, and Simberloff 1979; Ricklefs and Travis 1980; Joern and Lawlor 1980, 1981; Lawlor 1979, 1980b; Connor and Simberloff 1979; Simberloff and Boecklen 1981; Case and Sidell 1983; Schoener 1984). It has gradually become apparent that no algorithm for randomization is "perfectly random," but rather all such routines of necessity impose their own particular constraints. Various limitations of the technique are identified and discussed by Grant and Abbott (1980), Case (1983), Colwell and Winkler (1984), Gilpin and Diamond (1984), Harvey et al. (1983), Quinn and Dunham (1983), and Schoener (1984).

A biologically more realistic variant on the above technique consists of forming artificial communities by drawing species at random from a larger species pool. For example, with my data, pseudo-communities can be assembled consisting of various random subsets of real species taken from an overall pool of species consisting of all those found in all three continental desert-lizard systems (see Chapter 9). It is instructive to examine such psuedo-communities to see whether they differ from the real observed communities from which they are drawn.

Another intriguing approach to community ecology that has barely begun to be exploited involves more deliberate artificial "introduction," "removal," and/or "replacement." As one example of such an "experiment," a "resident" species in an empirically derived resource matrix can be systematically replaced with an "alien" (the same species as observed on another area). As a second example, a successful climbing gecko such as the Australian *Gehyra variegata* could be added to North American lizard community resource matrices. In either case, observed changes in various community-level macrodescriptors would then be compared with those of observed "control" communities.

For certain of these manipulations, data on relative abundances can be exploited to estimate carrying capacities, granted several simplifying assumptions (which may or may not be valid):

1. Species have reached dynamic equilibria with one another;
2. Overlap or similarity can be used to approximate "alpha";
3. The Lotka-Volterra competition equations hold.

Given these assumptions, carrying capacities ($K_i$) can be estimated by the sum of equilibrium population density ($N_i$) plus the summation of all the products of the appropriate "alpha" times the equilibrium population densities of each of the other species $\left(\sum_{j \neq i}^{n} \alpha_{ij} N_j\right)$. Note that such "carrying capacities" are not absolute but sampling-effort dependent. Once $K$'s have been calculated, one can examine predicted changes in population densities (increase, decrease, and to what extent) that will emerge from various hypothetical manipulations (see Chapter 9).

In the next two chapters, I use an empirical approach to search for answers to some fundamental questions about how natural communities are organized (Schoener 1974a asked some of these), including the following. How are available resources partitioned among members of ecological communities? How many, and which, niche dimensions separate species? Why? Is this niche segregation non random? If so, how? Does there seem to be a limit to how similar coexisting species can be? How much do pairs of species overlap? Does maximal tolerable niche overlap vary with various factors such as changing intensity of competition, the numbers of species, and/or environmental variability? If so, how? Is there complementarity in niche overlap along different niche dimensions? Are there clusters of functionally similar species with similar ecologies ("guilds"), separated from other such groups of species by lower overlaps? What are the effects of guild structure on community organization? To what extent in what ways do observed communities deviate from random models? How do randomly constructed pseudo-communities of real species differ from observed real communities? Do "alien" species "fit" into real communities less well than "resident" species did (e.g., do they equilibrate at lower densities)?

# 8 Empirical Results: Community Organization

In all three continental desert-lizard systems, a few species are disproportionately common, whereas the majority of species are relatively rare. As shown in Figures 3.1 and 4.5, rare species include the entire spectrum from extreme specialists to generalists. Most abundant species have moderate niche breadths. For the most part, species that are uncommon on one study area are also rare on other study sites (see Appendix A for data). These rare species raise a major unresolved problem: How do they manage to continue to exist at such precariously low densities without going extinct? (Of course, they may not actually be quite as rare as they appear to be.)

Uncommon species pose special problems of analysis: samples are frequently too small to characterize resource utilization patterns adequately, particularly at any given study area. One solution, taken by Inger and Colwell (1977), is simply to omit rare species altogether. However, I am most reluctant to adopt this methodology. These uncommon species are definitely present and doubtlessly influence other species. Even if the per species effects were negligible, the collective impact of many rare species on community structure could nevertheless be substantial, as envisioned in the notion of "diffuse competition" (MacArthur 1972). An alternative approach, used here, is to include rare species, lumping all specimens of each such species from all study sites. This procedure maximizes sample sizes and hence confidence in estimates of resource utilization patterns. Such a "merged deck" analysis essentially treats each continental desert-lizard system as a single site and, of necessity, obscures interesting area-to-area variation. Area-by-area analyses were also performed and generally gave comparable results. Both sorts of analyses are clearly useful since the most accurate reflection of reality must lie somewhere between them.

Relative abundances of lizard species are approximately log-normally distributed (Figure 8.1). The disparity between abundant versus uncommon species is greatest in North America, where just three species (*Cnemidophorus, Uta*, and *Callisaurus*) constitute over 80% of the total (Figure 8.1). In the Kalahari, species abundances are somewhat more equitably distributed: abundances are in excess of 5% in eight of the 22 species (about one-third); together these eight species comprise 75% of the total Kalahari saurofauna. In the Great Victoria desert, most species are uncommon; abundances of 55 of the 60 species (92%) are below 5%. Taken together, all 55 of these "rare" species represent fewer individuals than the five other common species, which sum to more than half the total. The evenness of relative abundances, or the "equitability" component of diversity (Tramer 1969), is usually estimated by the fraction of observed diversity over that maximally possible. Thus estimated, for the merged decks, equitability is lowest in Australia (.216), intermediate in North America (.263), and highest in the Kalahari (.491). In a site-by-site analysis, equitability is also lower in Australia (mean = .33, st.dev. = .13) than in North America (mean = .43, st.dev. = .19) or in the Kalahari (mean = .42, st.dev. = .10).

Lizard species diversity on these sites can be estimated using the relative abundances of various species in my collections as $p_i$'s in the diversity index of Simpson (1949). These estimates (Appendix A) are the best available, although they are clearly biased to the extent that

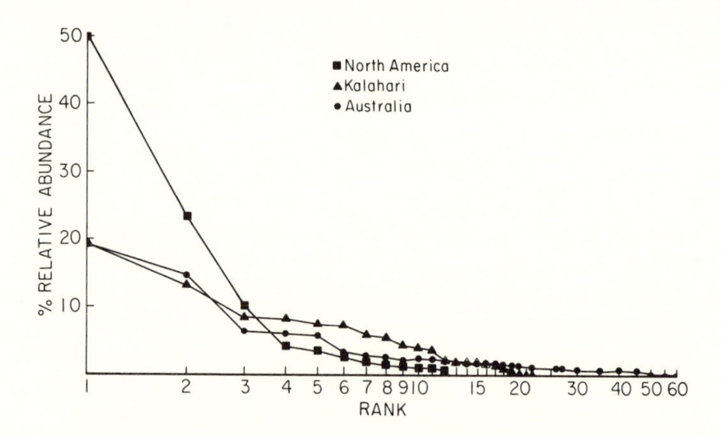


FIGURE 8.1. Estimated relative abundances of lizard species in each of the three continental desert systems, based on all specimens of each species collected. Equitability is highest in the Kalahari and lowest in Australia, where the majority of species are rare.

different species were probably not collected in proportion to their true relative abundances. In any case, such estimates of lizard diversity correlate strongly with the simpler measure of the number of lizard species ($r$ = .83), and, as noted in Chapter 2, are weakly correlated with the standard deviation in annual precipitation ($r$ = .47, $P$ < .05).

To facilitate examination of patterns of resource partitioning among these lizard assemblages, it is convenient to rank food resource states from those used most heavily by an entire saurofauna to those used least. Food resources are thus ranked on the basis of the row totals of an appropriate "merged deck" resource matrix.[1] As explained in Chapter 7, if these values are viewed as bioassays of prey availabilities, they can be used along with data on resource utilization (Appendices C and E) to estimate species-specific "electivities" on each of the various food resource states. Even though many species of lizards use abundant prey types, relatively few show high electivities for these resources (Figures 8.2, 8.3a,b, and 8.4a–e). Electivities on uncommon resources are, however, often high. Interestingly enough, most lizard species show high electivity for one or more foods, the identities of which tend to differ between consumer species. An assemblage of consumer species can be viewed as being somewhat analogous to a gearbox, with the electivities of various species representing the "cogs" meshing more-or-less neatly together.[2] In the next chapter, real assemblages are compared with various sorts of "pseudo-communities" in an attempt to ascertain just how good such fits among sympatric consumers actually are.

Frequency distributions of proportional utilization coefficients (merged decks) are remarkably similar for both microhabitat and diet, with a heavy preponderance of zero or very small elements and only a very few large ones (Table 8.1). Moreover, these values are also substantially alike among continental desert-lizard systems. The fit to a decaying exponential is fairly good. Electivities, estimated by standardizing proportional utilization entries in the resource matrix by division by row totals (resource "availabilities") and then readjusting

[1] Ranking discrete prey types in this way may actually reveal hidden but innate resource similarities and thus help to generate a somewhat "natural" resource spectrum, which in turn can be used to elucidate underlying biology. For example, most arboreal lizards tend to show high electivities for climbing and/or flying insects (wasps, flies, lepidopterans, mantids-phasmids, homoptera-hemiptera). High electivities tend to be clumped together when resource states are ranked in this manner.

[2] This analogy is dangerous and must not be pursued too far, for communities are not necessarily assembled for orderly and efficient function like a gearbox is, but rather each species of consumer may behave and evolve antagonistically toward the other members of its community.

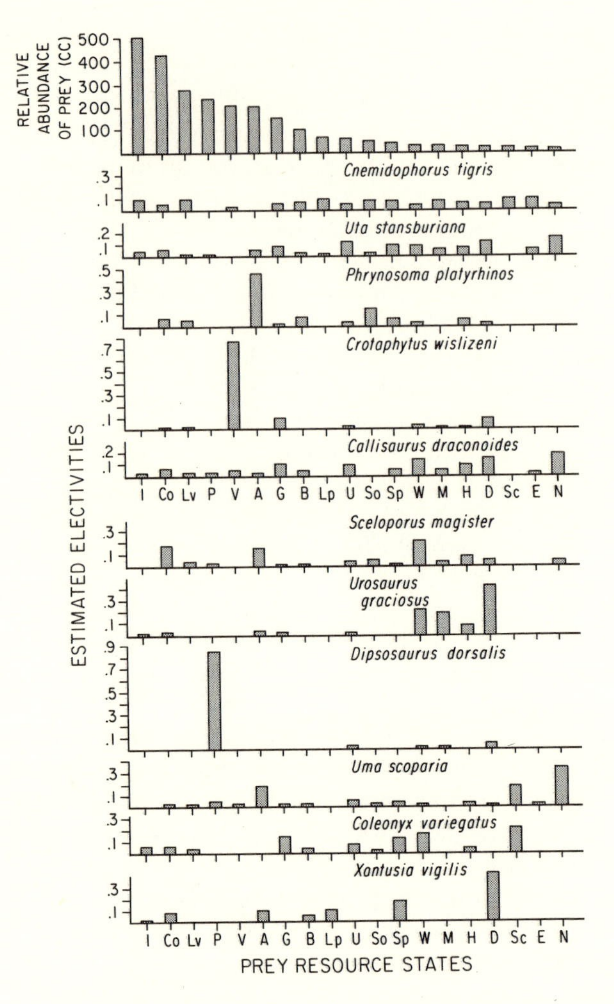


FIGURE 8.2. Patterns of food utilization among North American desert lizards, based on the merged deck for southern study sites. The top panel represents the overall diet by volume of the entire saurofauna, with prey types ranked from those used most to those used least (a crude bioassay of food "availabilities"). This same ranking is preserved in the panels below, which plot estimated electivities for each food resource state by each species of consumer.

ABBREVIATIONS FOR PREY TYPES IN FIGURES 8.2, 8.3, AND 8.4

| | | | |
|---|---|---|---|
| Ce | Centipedes | H | Bugs (Hemiptera and Homoptera) |
| Sp | Spiders | D | Flies (Diptera) |
| Sc | Scorpions | Lp | Butterflies and moths (Lepidoptera) |
| So | Solpugids (absent from Australia) | E | Insect eggs and pupae |
| A | Ants | | |

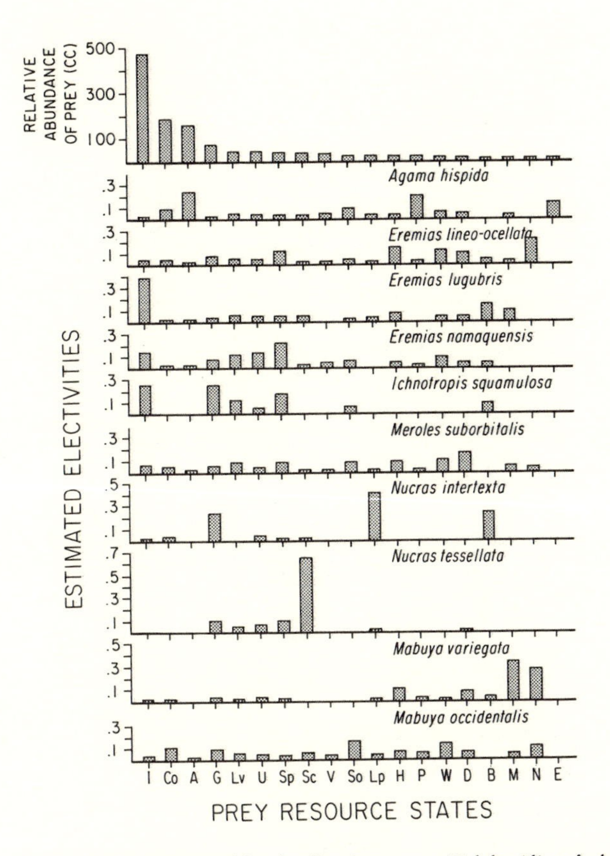


FIGURE 8.3a. Patterns of food utilization among Kalahari lizards, based on the merged deck for all study sites. The upper panel illustrates the overall diet by volume of the entire lizard fauna, with prey categories ranked from those most used to those least used (a crude bioassay of food "availabilities"). This same ranking is followed in the lower panels and in Figure 8.3b, which show estimated electivities for each food resource state by each species of consumer.

ABBREVIATIONS (*cont.*)

W Wasps and other non-ant hymenopterans
G Grasshopper and crickets
B Roaches (Blattids)
M Mantids and phasmids
N Adult Neuroptera (ant lions)
Co Beetles (Coleoptera)
I Termites (Isoptera)
Lv All insect larvae
U Miscellaneous arthropods, including unidentified items
V All vertebrate material, including sloughed lizard skins
P Plant materials (floral and vegetative)

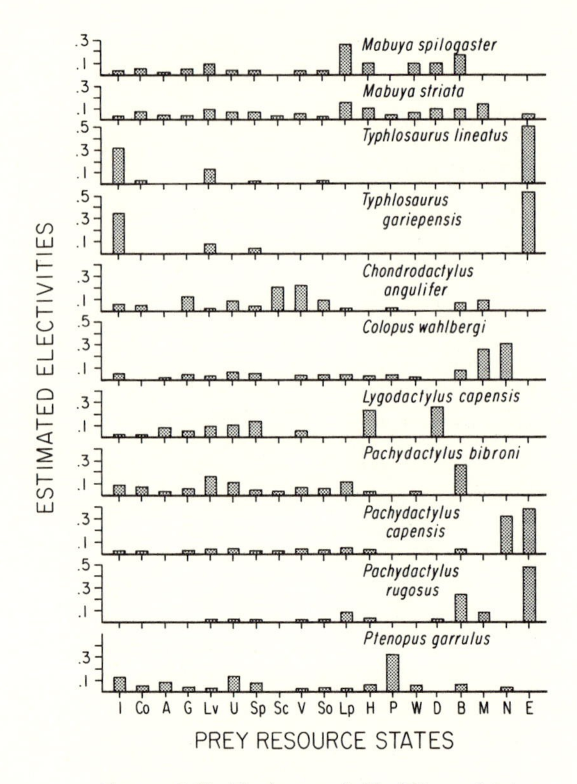


FIGURE 8.3b. The bottom half of Figure 8.3 (see legend for Figure 8.3a).

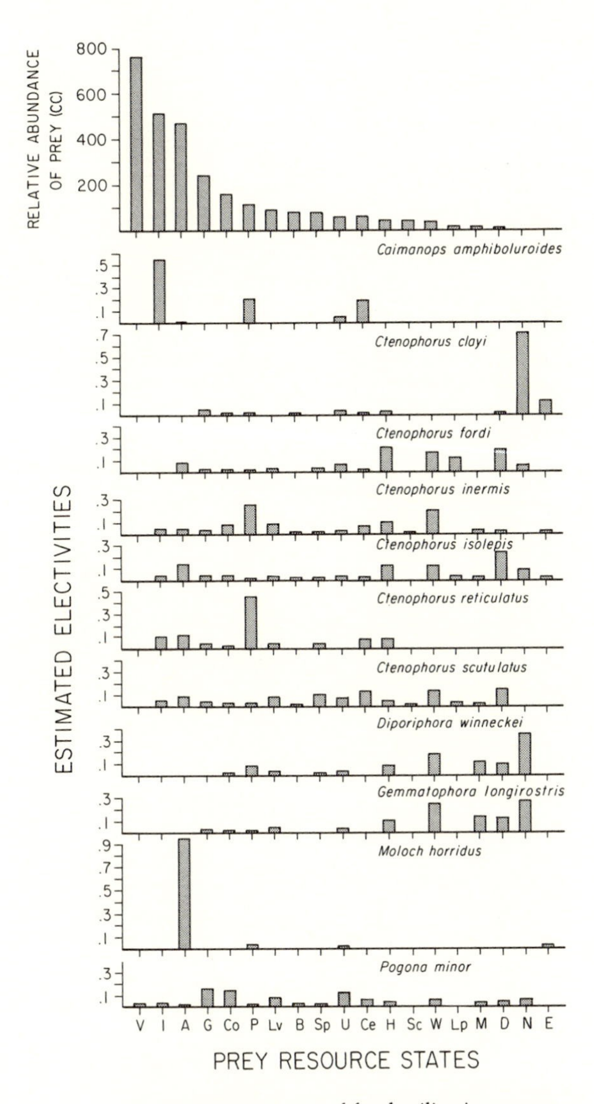


FIGURE 8.4a. Patterns of food utilization among Australian desert lizards, based on the merged deck for all study areas. The top panel represents the overall diet by volume of the entire saurofauna, with prey resource states ranked from those used most heavily to those used least (a crude bioassay of food "availabilities"). This same ranking is preserved in lower panels and in Figures 8.4b, 8.4c, 8.4d, and 8.4e, which portray estimated electivities for each food type by various consumer species.

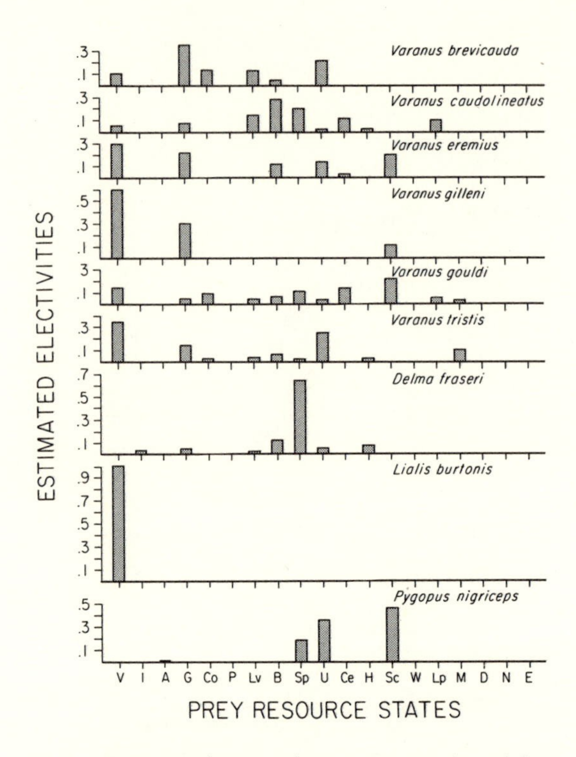


FIGURE 8.4b. Part of Figure 8.4 (see legend for Figure 8.4a).

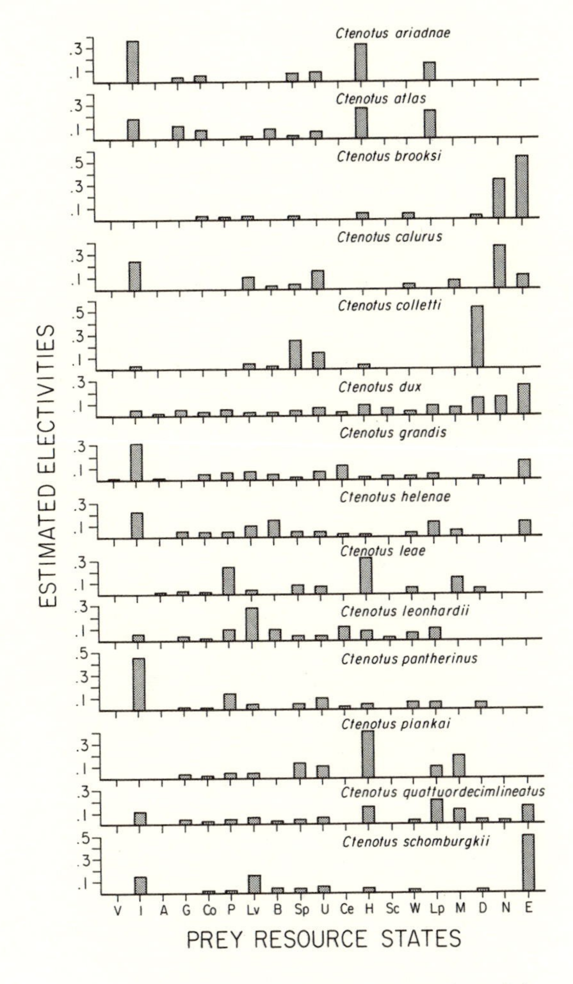


FIGURE 8.4c. Part of Figure 8.4 (see legend for Figure 8.4a).

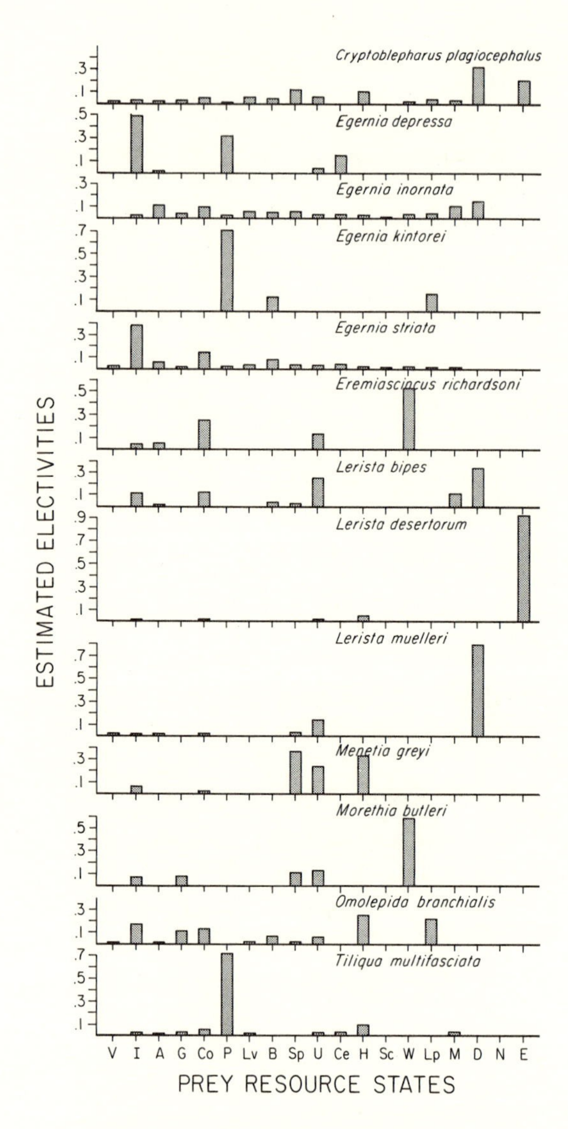


FIGURE 8.4d. Part of Figure 8.4 (see legend for Figure 8.4a).

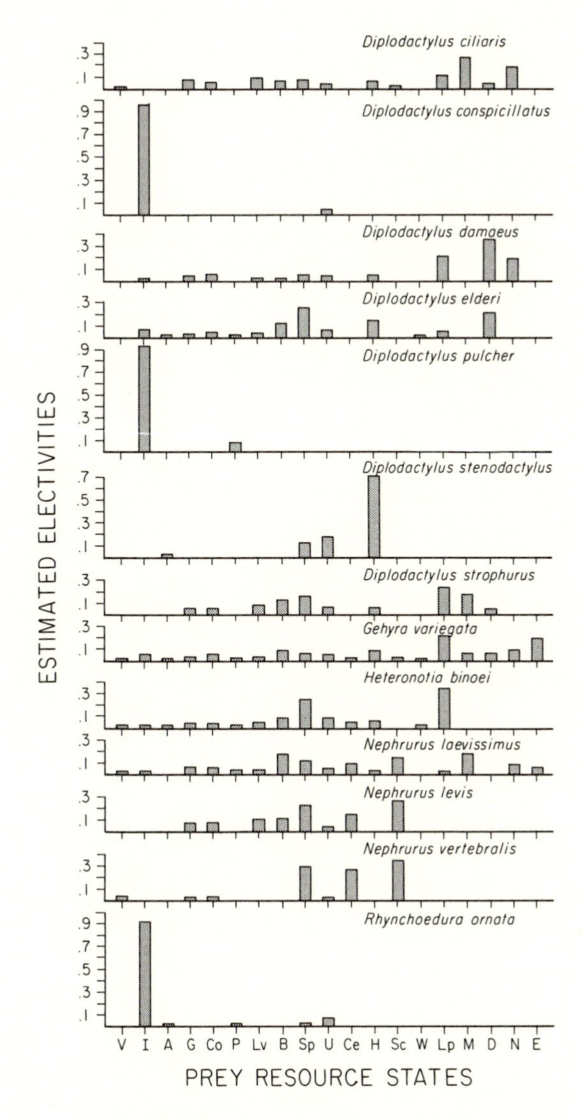


FIGURE 8.4e. Part of Figure 8.4 (see legend for Figure 8.4a).

TABLE 8.1. Frequency distributions of proportional resource utilization coefficients among all specimens of each species in each desert-lizard system (in percentages).

| | MICROHABITATS | | | PREY CATEGORIES | | |
|---|---|---|---|---|---|---|
| $p_i$ | *North America* | *Kalahari* | *Australia* | *North America* | *Kalahari* | *Australia* |
| 0.00 | | | | | | |
| | 81.8 | 74.6 | 75.6 | 76.1 | 77.7 | 77.2 |
| 0.05 | | | | | | |
| | 7.9 | 5.7 | 6.9 | 11.0 | 9.8 | 9.3 |
| 0.10 | | | | | | |
| | — | 3.8 | 3.4 | 2.4 | 5.0 | 4.4 |
| 0.15 | | | | | | |
| | — | 3.2 | 3.2 | 3.4 | 1.0 | 2.5 |
| 0.20 | | | | | | |
| | 0.6 | 3.8 | 1.7 | 1.9 | 1.5 | 1.1 |
| 0.25 | | | | | | |
| | — | 1.6 | 1.6 | 1.4 | 0.5 | 1.2 |
| 0.30 | | | | | | |
| | 0.6 | 1.9 | 1.4 | 0.5 | 0.8 | 0.7 |
| 0.35 | | | | | | |
| | 2.4 | 1.0 | 1.6 | 1.0 | 0.3 | 0.9 |
| 0.40 | | | | | | |
| | 1.2 | 1.3 | 0.6 | — | 0.3 | 0.3 |
| 0.45 | | | | | | |
| | 2.4 | 1.0 | 0.3 | — | 0.5 | 0.2 |
| 0.50 | | | | | | |
| | — | 0.6 | 1.2 | 1.0 | 0.8 | 0.3 |
| 0.55 | | | | | | |
| | — | 0.6 | 0.1 | 0.5 | — | 0.4 |
| 0.60 | | | | | | |
| | — | 0.3 | 0.3 | — | 0.5 | 0.1 |
| 0.65 | | | | | | |
| | — | — | 0.4 | — | 0.3 | 0.2 |
| 0.70 | | | | | | |
| | — | — | 0.6 | — | 0.3 | 0.3 |
| 0.75 | | | | | | |
| | — | — | 0.1 | — | 0.3 | 0.1 |
| 0.80 | | | | | | |
| | 1.2 | — | 0.5 | — | — | 0.4 |
| 0.85 | | | | | | |
| | 0.6 | — | 0.1 | 0.5 | — | 0.3 |
| 0.90 | | | | | | |
| | — | — | 0.3 | — | 0.5 | — |
| 1.00 | | | | | | |
| | 1.2 | 0.6 | — | 0.5 | 0.3 | 0.4 |
| Total number | 154 | 315 | 900 | 209 | 399 | 1140 |

the coefficients for each species so that they sum to unity, are distributed similarly (Table 8.2).

It is virtually impossible to evaluate the degree of interdependence of food and microhabitat niche dimensions in most lizard species, due to their mobility. (This problem applies both to utilizations and to electivities.) However, in the relatively sedentary Kalahari subterranean skink *Typhlosaurus lineatus*, which specializes on fairly sedentary prey (termites), the degree to which foods that are eaten vary with microhabitat of collection has been assessed (Huey et al. 1974; Pianka et al. 1979). Most species and castes of termites are eaten in fairly similar proportions by *Typhlosaurus* collected in different microhabitats (Table 8.3), indicating that these two niche dimensions are largely independent, at least in this lizard species.

TABLE 8.2. Frequency distributions of electivities among all specimens of each species in each continental desert-lizard system for microhabitat and food resources (percentages).

| | MICROHABITATS | | | PREY CATEGORIES | | |
|---|---|---|---|---|---|---|
| $a_i$ | *North America* | *Kalahari* | *Australia* | *North America* | *Kalahari* | *Australia* |
| 0.00 | 70.1 | 68.9 | 77.0 | 68.4 | 69.4 | 75.8 |
| 0.05 | 4.6 | 7.9 | 4.9 | 18.2 | 16.0 | 10.5 |
| 0.10 | 5.2 | 8.6 | 4.1 | 5.3 | 6.3 | 3.7 |
| 0.15 | 9.7 | 3.8 | 2.7 | 3.8 | 1.8 | 2.6 |
| 0.20 | 2.6 | 2.2 | 1.7 | 1.4 | 2.0 | 2.1 |
| 0.25 | 1.9 | 3.2 | 1.9 | — | 1.5 | 0.9 |
| 0.30 | 1.9 | 1.3 | 1.0 | 0.5 | 1.0 | 0.6 |
| 0.35 | 1.3 | 1.3 | 1.8 | — | 0.8 | 0.7 |
| 0.40 | 0.7 | 0.3 | 0.9 | 1.0 | — | 0.6 |
| 0.45 | — | 0.3 | 0.8 | 0.5 | 0.5 | 0.5 |
| 0.50 | 0.7 | 1.0 | 0.6 | — | 0.3 | 0.4 |
| 0.55 | — | 0.6 | 0.7 | — | — | 0.3 |
| 0.60 | — | — | 0.3 | — | 0.3 | 0.3 |
| 0.65 | — | — | — | — | — | 0.3 |
| 0.70 | — | — | 0.3 | 0.5 | — | — |
| 0.75 | — | — | 0.1 | — | — | 0.1 |
| 0.80 | — | — | 0.1 | — | 0.3 | 0.1 |
| 0.85 | — | — | 0.2 | 0.5 | — | — |
| 0.90 | — | — | 0.2 | — | — | 0.1 |
| 0.95 | 1.3 | 0.6 | 0.7 | — | — | 0.5 |
| 1.00 | | | | | | |
| Total number | 154 | 315 | 900 | 209 | 399 | 1140 |

The degree of interdependence of various resource states should also be reflected in correlations in (1) proportional utilization coefficients among resource states over all consumers, as well as in (2) the overall utilizations among resources over different study areas. The vast majority of such correlation coefficients, including those *between* microhabitat and prey resources, are weak (Table 8.4), suggesting a fairly substantial degree of independence among various resource states. Both the *Typhlosaurus* results mentioned above and these correlations imply that the products of various food and microhabitat niche metrics are the most appropriate estimators of overall patterns (see also May 1975). In addition, this high degree of independence among resource states dictates that ecological niche space has high dimensionality,

TABLE 8.3. Percentage and total number of prey items eaten by *Typhlosaurus lineatus* under specific microhabitats.

| | LOGS | | LEAF LITTER | | CROTALARIA | | GRASS |
|---|---|---|---|---|---|---|---|
| | *Sandplain Areas* | *Sandridge Areas* | *Sandplain Areas* | *Sandridge Areas* | *Sandplain Areas* | *Sandridge Areas* | *Sandridge Areas* |
| *Allodontermes* (*schultzei*?) | | | | | | | |
| Minor workers | 32.3 | 34.2 | 34.4 | 21.9 | 29.2 | 30.2 | 40.7 |
| Major workers | 24.5 | 52.7 | 48.0 | 74.1 | 42.3 | 64.5 | 54.8 |
| *Psammotermes allocerus* | | | | | | | |
| Workers | 36.5 | 7.1 | 7.3 | 1.8 | 24.6 | 1.9 | 1.6 |
| Soldiers | 1.0 | 3.9 | 3.0 | 1.4 | 3.2 | 1.4 | 1.8 |
| *Hodotermes mossambicus* | 5.5 | 1.8 | 6.0 | 0.6 | 0.7 | 1.1 | 0.7 |
| Other termites | 0.3 | 0.3 | 1.4 | 0.1 | — | 0.9 | 0.4 |
| Total number of termites | 2385 | 4214 | 2500 | 4746 | 1037 | 3386 | 2620 |

which unfortunately renders multivariate techniques of little utility in reducing dimensionality (these same techniques work splendidly on highly correlated data sets such as morphometrics; see also Chapter 11).

Frequency distributions of observed overlap are similar for both the merged decks and in area-by-area analyses (Tables 8.5, 8.6, and 8.7), tending to be skewed toward low values. In the Kalahari, heavy consumption of termites results in somewhat higher dietary overlap values. Similarities, computed using estimated electivities rather than proportional utilizations (Lawlor 1980a), are distributed similarly (Table 8.8) and are positively correlated with overlaps (Table 8.9). Again the Kalahari diet is aberrant, showing the weakest correlation.

For the merged decks, plots of standard deviation in overlap versus nearness rank in niche space show fairly convincing humps in Australia (diet, microhabitat, and product overlaps) as well as for diet in the Kalahari (see Figures 9.1 through 9.6). Area-by-area analyses also reveal numerous humps in such plots of standard deviation in dietary and/or microhabitat overlap versus nearness rank in niche space in both the Kalahari and Australian lizard faunas (Figures 8.6 and 8.7). Such humps are not as evident in the less diverse North American lizards (Figure 8.5), implying that species are not as clustered in niche space ("guild structure" is reduced).

TABLE 8.4. Frequency distributions of correlation coefficients between various resource states ($p_i$'s) over all consumers in each continental desert-lizard system (based on merged decks).

| | NORTH AMERICA | | | KALAHARI | | | AUSTRALIA | | |
|---|---|---|---|---|---|---|---|---|---|
| *r* | *Micro-habitat* | *Diet* | *Micro-habitat vs. Diet* | *Micro-habitat* | *Diet* | *Micro-habitat vs. Diet* | *Micro-habitat* | *Diet* | *Micro-habitat vs. Diet* |
| +1.0 | 2 | | | 1 | 1 | | | | |
| + .9 | 5 | 4 | 1 | 2 | 1 | | | | |
| + .8 | 3 | 3 | 4 | 2 | 1 | | 1 | | |
| + .7 | 6 | 5 | 7 | 1 | 4 | 3 | | | |
| + .6 | 1 | 6 | 8 | 3 | 5 | 9 | | | |
| + .5 | 0 | 1 | 11 | 5 | 8 | 14 | 1 | 2 | 1 |
| + .4 | 4 | 8 | 15 | 4 | 10 | 15 | 1 | 7 | 8 |
| + .3 | 3 | 12 | 9 | 5 | 8 | 19 | 3 | 11 | 10 |
| + .2 | 3 | 12 | 21 | 4 | 20 | 20 | 6 | 15 | 21 |
| + .1 | 4 | 22 | 24 | 10 | 25 | 30 | 11 | 21 | 45 |
| 0.0 | 3 | 14 | 19 | 6 | 26 | 39 | 19 | 36 | 70 |
| − .1 | 15 | 19 | 34 | 9 | 25 | 64 | 34 | 47 | 107 |
| − .2 | 12 | 32 | 59 | 25 | 22 | 42 | 21 | 26 | 20 |
| − .3 | 13 | 20 | 29 | 19 | 5 | 20 | 6 | 5 | 3 |
| − .4 | 13 | 8 | 14 | 8 | 6 | 9 | 2 | 1 | |
| − .5 | 4 | 4 | 8 | 1 | 2 | 1 | | | |
| − .6 | | 1 | 2 | | 2 | | | | |
| − .7 | | | 1 | | | | | | |

TABLE 8.5. Frequency distributions of dietary overlap for merged decks and in area-by-area analyses in each continental desert-lizard system (in percentages).

| Overlap Value | North America | | Kalahari | | Australia | |
|---|---|---|---|---|---|---|
| | *Merged* | *Area-by-Area* | *Merged* | *Area-by-Area* | *Merged* | *Area-by-Area* |
| 0.00 | 14.6 | 9.9 | — | 5.9 | 15.5 | 33.7 |
| .05 | 10.9 | 11.3 | 0.5 | 4.5 | 12.0 | 11.5 |
| .10 | 5.5 | 3.8 | 1.9 | 3.5 | 7.2 | 7.4 |
| .15 | 3.6 | 5.6 | 3.3 | 3.3 | 5.8 | 5.3 |
| .20 | 1.8 | 3.8 | 2.4 | 3.6 | 5.4 | 4.6 |
| .25 | 5.5 | 5.2 | 6.2 | 3.7 | 5.3 | 3.6 |
| .30 | 1.8 | 3.3 | 4.8 | 4.7 | 5.3 | 3.4 |
| .35 | 1.8 | 4.2 | 4.8 | 4.0 | 4.8 | 2.6 |
| .40 | 3.6 | 4.7 | 3.8 | 3.2 | 4.1 | 2.6 |
| .45 | — | 5.2 | 5.2 | 2.2 | 4.0 | 3.2 |
| .50 | 3.6 | 4.2 | 4.3 | 3.0 | 3.0 | 2.3 |
| .55 | 7.3 | 3.8 | 5.7 | 4.2 | 2.7 | 2.1 |
| .60 | 5.5 | 4.7 | 3.3 | 3.4 | 2.5 | 2.6 |
| .65 | 3.6 | 4.7 | 1.9 | 4.1 | 2.8 | 1.8 |
| .70 | 5.5 | 6.1 | 2.4 | 6.7 | 2.3 | 1.7 |
| .75 | 9.1 | 4.2 | 5.7 | 6.2 | 2.8 | 1.8 |
| .80 | 3.6 | 3.3 | 5.7 | 4.2 | 2.2 | 1.3 |
| .85 | 5.5 | 3.3 | 4.3 | 7.4 | 3.3 | 1.5 |
| .90 | 3.6 | 2.8 | 10.5 | 4.5 | 2.2 | 1.5 |
| .95 | 3.6 | 5.6 | 15.7 | 9.4 | 3.2 | 1.9 |
| 1.00 | — | 0.5 | 7.6 | 8.0 | 3.7 | 4.1 |
| Total number of pairs | 55 | 213 | 210 | 993 | 1770 | 3760 |
| Mean | .43 | .42 | .65 | .57 | .35 | .26 |
| Standard deviation | .33 | .31 | .28 | .32 | .31 | .31 |

TABLE 8.6. Frequency distributions of microhabitat overlap for merged decks and in area-by-area analyses in each continental desert-lizard system (in percentages).

| OVERLAP VALUE | NORTH AMERICA | | KALAHARI | | AUSTRALIA | |
|---|---|---|---|---|---|---|
| | *Merged* | *Area-by-Area* | *Merged* | *Area-by-Area* | *Merged* | *Area-by-Area* |
| 0.00 | 52.7 | 27.2 | 28.1 | 39.7 | 17.5 | 33.8 |
| .05 | 7.3 | 7.5 | 5.7 | 5.3 | 13.6 | 8.5 |
| .10 | — | 3.8 | 10.0 | 5.6 | 11.4 | 7.1 |
| .15 | — | 2.4 | 8.1 | 3.9 | 6.6 | 4.2 |
| .20 | — | 0.9 | 8.1 | 4.8 | 6.3 | 3.9 |
| .25 | — | 0.5 | 4.3 | 3.2 | 3.6 | 2.9 |
| .30 | — | — | 2.9 | 2.2 | 3.6 | 3.8 |
| .35 | — | 0.9 | 3.3 | 3.2 | 3.2 | 3.2 |
| .40 | — | 2.4 | 2.4 | 2.5 | 2.4 | 2.8 |
| .45 | 3.6 | 4.2 | 3.8 | 2.7 | 2.8 | 2.6 |
| .50 | 1.8 | 3.3 | 0.5 | 2.8 | 3.0 | 3.1 |
| .55 | — | 1.9 | 2.9 | 2.9 | 4.1 | 3.9 |
| .60 | — | 2.4 | 1.0 | 2.3 | 2.9 | 2.2 |
| .65 | — | 1.9 | 1.9 | 2.3 | 3.2 | 2.2 |
| .70 | — | 2.4 | 2.4 | 1.8 | 2.6 | 3.2 |
| .75 | — | 5.2 | 1.4 | 2.1 | 2.2 | 2.2 |
| .80 | 14.6 | 6.6 | 0.5 | 2.2 | 2.7 | 2.3 |
| .85 | 1.8 | 5.2 | 2.4 | 2.7 | 2.2 | 2.0 |
| .90 | 5.5 | 8.9 | 3.3 | 2.6 | 2.0 | 2.1 |
| .95 | 1.8 | 6.1 | 5.2 | 2.4 | 3.1 | 2.4 |
| 1.00 | 10.9 | 6.6 | 1.9 | 2.4 | 1.2 | 1.8 |
| Total number of pairs | 55 | 213 | 210 | 993 | 1770 | 3760 |
| Mean | .34 | .46 | .29 | .27 | .31 | .28 |
| Standard deviation | .42 | .39 | .32 | .32 | .30 | .31 |

Table 8.7. Frequency distributions of product overlaps (diet × microhabit) for merged decks and in area-by-area analyses in each continental desert-lizard system (in percentages).

| Overlap Value | North America | | Kalahari | | Australia | |
|---|---|---|---|---|---|---|
| | *Merged* | *Area-by-Area* | *Merged* | *Area-by-Area* | *Merged* | *Area-by-Area* |
| 0.00 | 67.3 | 41.8 | 32.4 | 46.9 | 51.2 | 66.3 |
| .05 | 7.3 | 13.2 | 16.2 | 10.4 | 17.0 | 10.6 |
| .10 | 3.6 | 5.2 | 8.6 | 7.6 | 7.1 | 5.5 |
| .15 | — | 5.6 | 9.5 | 5.7 | 6.0 | 3.4 |
| .20 | 7.3 | 5.6 | 4.8 | 4.2 | 3.1 | 2.6 |
| .25 | 1.8 | 1.9 | 4.3 | 3.9 | 1.8 | 2.3 |
| .30 | 1.8 | 2.8 | 5.2 | 3.9 | 2.2 | 1.5 |
| .35 | 1.8 | 1.9 | 2.9 | 2.0 | 1.8 | 1.1 |
| .40 | 1.8 | 3.3 | 2.9 | 1.8 | 1.5 | 0.9 |
| .45 | — | 2.8 | 3.3 | 1.8 | 1.6 | 1.0 |
| .50 | 1.8 | 0.9 | 1.0 | 1.2 | 1.1 | 0.8 |
| .55 | — | 3.8 | 0.5 | 1.1 | 1.2 | 0.9 |
| .60 | — | 3.8 | 1.0 | 1.5 | 0.8 | 0.4 |
| .65 | — | 0.9 | 1.0 | 1.4 | 0.7 | 0.6 |
| .70 | — | 1.4 | 0.5 | 1.2 | 0.6 | 0.5 |
| .75 | 3.6 | 2.4 | — | 1.3 | 0.4 | 0.3 |
| .80 | — | — | 0.5 | 0.9 | 0.5 | 0.4 |
| .85 | — | 1.9 | 1.9 | 0.9 | 0.5 | 0.3 |
| .90 | — | 0.5 | 2.4 | 0.7 | 0.6 | 0.2 |
| .95 | 1.8 | 0.5 | 1.0 | 0.7 | 0.6 | 0.4 |
| 1.00 | — | — | 0.5 | 0.7 | — | 0.2 |
| Total number of pairs | 55 | 213 | 210 | 993 | 1770 | 3760 |
| Mean | .10 | .19 | .19 | .16 | .11 | .08 |
| Standard deviation | .21 | .25 | .24 | .24 | .19 | .17 |

TABLE 8.8. Frequency distributions of similarities, computed using electivities based on merged decks, in each continental desert-lizard system (in percentages).

| SIMILARITY VALUE | NORTH AMERICA | | KALAHARI | | AUSTRALIA | |
|---|---|---|---|---|---|---|
| | *Diet* | *Microhabitat* | *Diet* | *Microhabitat* | *Diet* | *Microhabitat* |
| .00 | 10.9 | 52.7 | 3.8 | 30.0 | 17.2 | 27.9 |
| .05 | 7.3 | 3.6 | 6.7 | 9.1 | 11.9 | 16.4 |
| .10 | 16.4 | — | 12.4 | 6.7 | 10.6 | 10.3 |
| .15 | 7.3 | — | 8.1 | 6.2 | 10.3 | 7.0 |
| .20 | 3.6 | 1.8 | 9.5 | 6.7 | 8.3 | 6.8 |
| .25 | 5.5 | 10.9 | 8.1 | 5.7 | 6.4 | 3.8 |
| .30 | 7.3 | — | 8.6 | 3.8 | 5.3 | 3.5 |
| .35 | 5.5 | 5.5 | 7.1 | 3.8 | 5.3 | 2.9 |
| .40 | 3.6 | 1.8 | 8.6 | 6.2 | 4.8 | 2.3 |
| .45 | 5.5 | — | 4.8 | 2.4 | 3.1 | 2.0 |
| .50 | 7.3 | — | 4.8 | 1.0 | 2.6 | 2.4 |
| .55 | 3.6 | 3.6 | 4.8 | 1.0 | 2.9 | 2.5 |
| .60 | 10.9 | 1.8 | 1.4 | 1.0 | 3.1 | 2.6 |
| .65 | 1.8 | 1.8 | 3.8 | 2.9 | 1.9 | 2.0 |
| .70 | 1.8 | 1.8 | 2.9 | 1.9 | 1.6 | 2.0 |
| .75 | — | 1.8 | 1.9 | 3.3 | 1.1 | 1.2 |
| .80 | — | — | 1.9 | 2.4 | 1.4 | 1.4 |
| .85 | — | 7.3 | — | 2.4 | 1.0 | 1.0 |
| .90 | — | 5.5 | 0.5 | 1.0 | 0.8 | 1.0 |
| .95 | 1.8 | — | — | 1.9 | 0.3 | 1.2 |
| 1.00 | — | — | 0.5 | 1.0 | 0.2 | 0.3 |
| Mean | .305 | .246 | .320 | .262 | .248 | .212 |
| Standard deviation | .232 | .321 | .213 | .288 | .229 | .249 |
| Total number of pairs | 55 | 55 | 210 | 210 | 1770 | 1770 |

TABLE 8.9. Correlations between overlaps based on $p_i$'s versus similarities computed using electivities, for merged decks.

| ASSEMBLAGE | DIET | MICROHABITAT |
|---|---|---|
| North America | .789* | .815* |
| Kalahari | .393ns | .926† |
| Australia | .550† | .763† |

ns = not significant  *$P < .01$  †$P < .001$

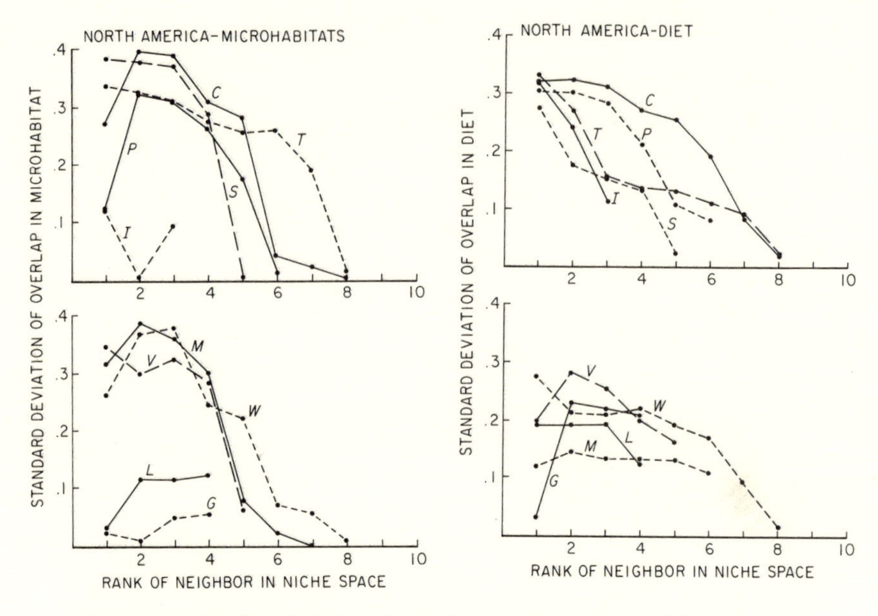


**Figure 8.5.** Standard deviations in overlap versus nearness rank in niche space for microhabitat (left) and diet (right) in North American desert lizards on various study areas (designated by letter codes).

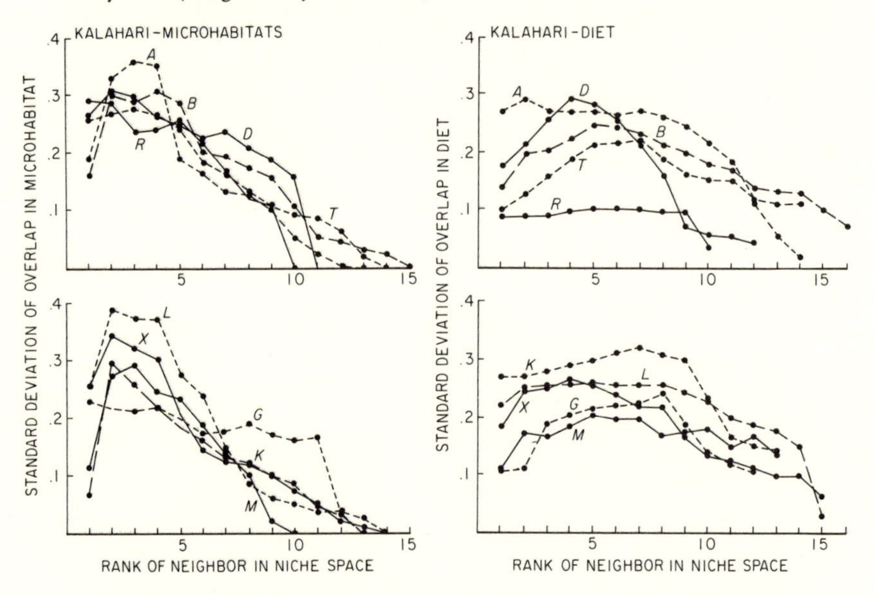


**Figure 8.6.** Standard deviations in overlap versus nearness rank in niche space for microhabitat (left) and diet (right) in Kalahari lizards on various study sites (designated by letter codes).

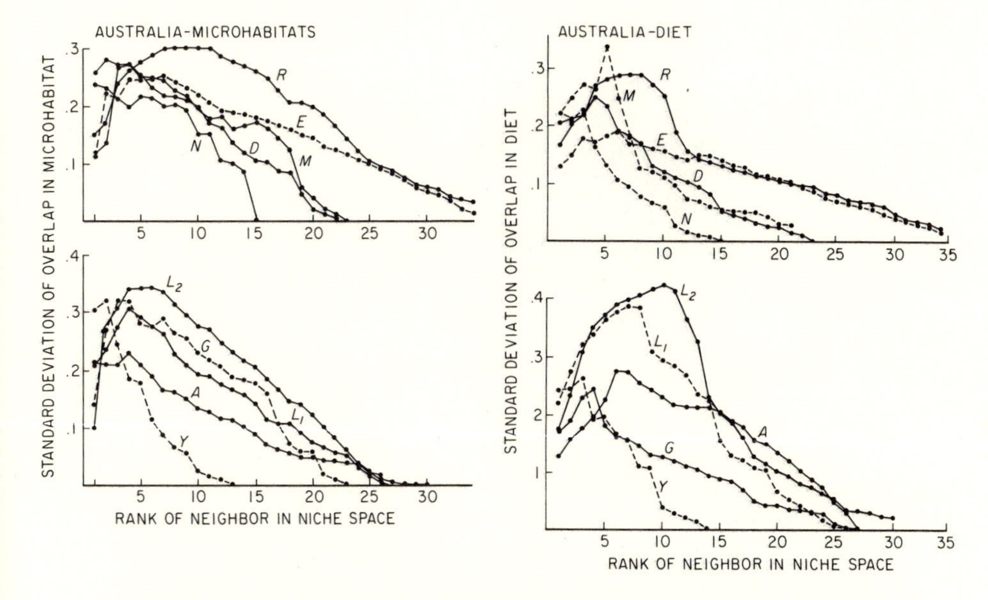


FIGURE 8.7. Standard deviations in overlap versus nearness rank in niche space for microhabitat (left) and diet (right) in Australian desert lizards on various study areas (designated by letter codes).

# 9 Games Computers Play: Pseudo-Communities

As indicated in Chapter 7, an extremely potent way to highlight structure in real natural communities is to compare them with various sorts of pseudo-communities generated on computers. This is an endless exercise in that an infinite[1] variety of potentially instructive pseudo-communities can be imagined.

As a background against which to compare the patterns and trends reported in Chapter 8, species lists for various subcommunities were drawn at random from the overall species pool composed of all 90-odd species in all three continental desert-lizard systems. Such subsets are innately biased toward the more speciose Australian system, which includes two-thirds of the total species. Even so, most such artificially constructed pseudo-communities include species from more than one continent. Interestingly, intercontinental assemblages are not devoid of pattern—for example, plots of standard deviation versus nearness rank display humps that change in an orderly way with number of species (the rank at which humps peak increases as number of species increases).

Another sort of pseudo-community consists of variously randomized replicates based on a particular real community as a prototype, using an observed list of species. Perhaps one of the least interesting but at the same time one of the most random ways to construct such a pseudo-community is to draw numbers from the uniform random distribution and assign these at random to serve as resource utilization coefficients in the resource matrix (Pianka et al. 1979; Joern and Lawlor 1980; Lawlor 1980b). Such randomly constructed communities (treatment 1)

[1] Just as one must beware of mindless descriptions and mindless experimentation, one should be wary of mindless computer simulations (we don't want to become PacMan's toys!).

have little structure of interest: (1) overlap is high and unimodal; (2) variance in overlap is low; and (3) guild structure is essentially nonexistent.

A similar but slightly more realistic randomization routine (treatment 2) again assigns utilization coefficients drawn from the uniform random distribution at random, but incorporates a little of the structure of the real community by retaining zero utilization coefficients for resource states that are not used among each consumer species. Still another randomization algorithm (treatment 3) randomly rearranges *observed* utilization coefficients (actual $p_i$'s) over all resource states. Finally, another process (treatment 4) interchanges *observed* utilization coefficients randomly, but only among the resource states actually used by each species. Both treatments 3 and 4 thus use observed $p_i$'s and retain observed niche breadths, whereas treatments 2 and 4 retain the zero structure of the resource matrix. Table 9.1 summarizes these four randomization algorithms.

For each randomization routine, a computer is used to generate 100 randomized replicates that are then compared with the original observed community. Distributions of overlap differ markedly from observed communities in treatments 1 and 2, but are much closer under treatments 3 and 4, which retain more of the original community's structure. Note that treatment 4 incorporates the most of the original community's organization and treatment 1 the least. Treatments 2 and 3 are intermediate, with one retaining niche breadths and the other the zero structure of the resource matrix. All four randomization algorithms simultaneously distort possible competitive avoidances and resource availabilities, at least when proportional utilization coefficients are used to characterize consumers. If, however, consumer utilization patterns are described by means of electivities (Lawlor 1980b), resource availabilities can, in effect, be "held constant." While statistics are difficult in comparing 100 cases against a single case, if 99 out of the 100 differ from the observed, one can be fairly confident that there is a significant difference.

Results of randomizing the observed communities (merged data sets for all the lizards in each continental desert-lizard system) with each of these four algorithms are shown in Figures 9.1–9.6. Interestingly enough, in treatment 1, standard deviations invariably increase monotonically as one moves from the nearest niche neighbor to the most distant (these randomized replicates of real communities have thus lost all of their guild structure). Another feature of the treatment 1 randomized pseudo-communities is that, at all ranks of nearness in niche space, average overlap is substantially higher than in observed

TABLE 9.1. Summary of four algorithms for randomization of resource matrices.

| | Zero Structure Destroyed | Zero Structure Retained |
|---|---|---|
| $p_i$'s from the uniform random distribution | Treatment 1 | Treatment 2 |
| observed $p_i$'s retained | Treatment 3 | Treatment 4 |

Treatment 1: Draws $p_i$'s from the uniform random distribution for all resource states for each species (this is most random).

Treatment 2: Similar to treatment 1, but slightly more realistic. Utilization coefficients of zero are retained (lizards cannot begin to use a new resource type), and observed utilization coefficients are replaced with those drawn from the uniform random distribution. Some of the structure of the real resource matrix is thus retained.

Treatment 3: Observed $p_i$'s are shuffled randomly over all resource states, including zero elements. Observed niche breadths are thus held constant.

Treatment 4: Observed $p_i$'s are rearranged randomly, but only among those resource states actually used in real community. Observed niche breadths are thus held constant and the zero structure of the resource matrix is retained.

NOTE: These procedures can be undertaken by using estimates of electivities, or $a_i$'s, rather than using simple utilization coefficients ($p_i$'s). This approach effectively holds resource "availabilities" constant.

communities. In North America and the Kalahari, treatment 2 randomized communities are similar to those of treatment 1 for the food niche dimension but not for microhabitats. This intriguing result suggests a fundamental difference between the food resource matrix and the resource matrix for microhabitat utilization: moreover, it implicates the zero structure of these two matrices as being fundamentally different (clearly this observation forces one to confront the original data once again, but armed with new insights as to what to look for in terms of pattern). In Australia, treatment 2 pseudo-communities differ markedly from treatment 1 pseudo-communities for both diet and microhabitat (in fact, treatment 2 pseudo-communities look much more like the observed community); this, of course, again implicates the zero structure of these two resource matrices as being critical in structuring the real communities.

As would be expected, treatments 3 and 4 tend to produce pseudo-

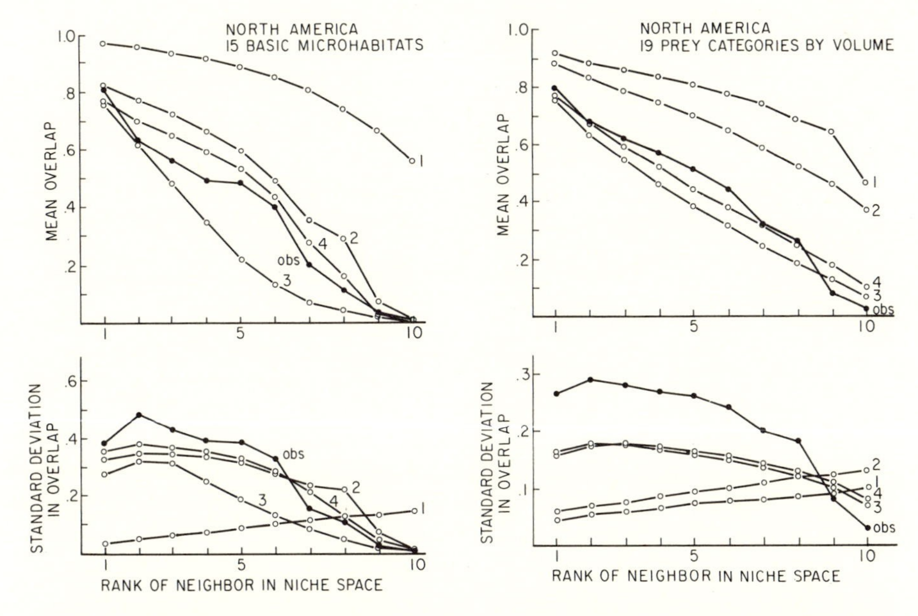


FIGURE 9.1. Plots of mean (upper panel) and standard deviation (lower panel) in overlap versus nearness rank in niche space for microhabitat resources in North America. Each randomization algorithm was performed on 100 randomized pseudo-communities and the average of all are shown here (variance between these replicates is slight—standard errors on the curves for randomized pseudo-communities are extremely low; if plotted, most would encompass a belt about a millimeter wide).

FIGURE 9.2. Plots like those shown in Figure 9.1, but for prey resources in North America. Standard errors around the lines plotted for randomized pseudo-communities are very small.

communities that are more similar to observed communities, although there are some consistent differences. All four randomization routines tend to destroy some of the original community's guild structure in that the standard deviation versus nearness rank curves become flatter. Standard deviation curves of overlap versus nearness rank for the randomized pseudo-communities also smooth out the "bumps" in plots for observed communities; this seems to be more than merely an averaging effect—for example, for the Kalahari food niche dimension, the second bump in the standard deviation curve from about 10 to 15 species out in niche space is destroyed (Figure 9.4). Also, for the Kalahari microhabitat dimension, notice that all randomizations

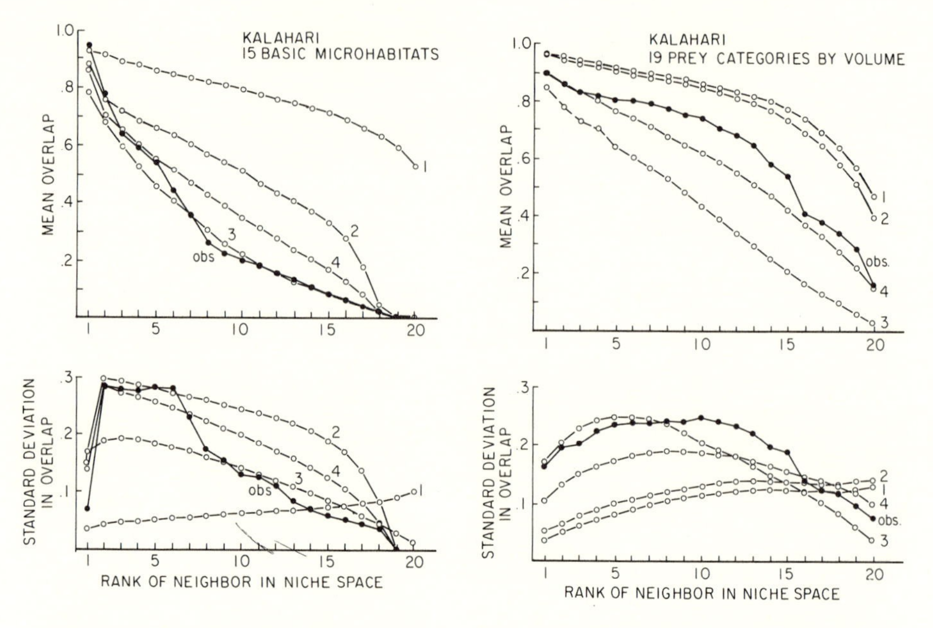


FIGURE 9.3. Plots like those shown in Figure 9.1, but for Kalahari lizards (microhabitat resources).

FIGURE 9.4. Plots like those shown in Figure 9.1, but for food resources in Kalahari lizards.

tend to overestimate the variance in niche overlap for distant neighbors (Figure 9.3). All these results underscore the fact that species are indeed clustered in resource space; these saurofaunas are not assembled randomly but possess substantial guild structure.

How well do the resource utilization patterns observed among sympatric species on a given study site "fit" together? Can evidence be found for ecological adjustments among coexisting species? In an effort to address these questions, a fairly extensive series of artificial "removal-introduction experiments"[2] were undertaken. Area-by-area resource matrices for diet were assembled and analyzed to estimate the "electivities" of each consumer species on each prey resource state, as outlined in Chapter 7. Each "resident" species on every study site was then systematically replaced by the same species as it was actually observed on each of the other study areas where that species occurred naturally (these are termed "aliens"). A moderately large number of such "transplants" can be made: for a ubiquitous species found on ten study areas, nine alien "introductions" are possible on each site,

[2] L. R. Lawlor's FORTRAN wizardry made this massive transplant analysis possible.

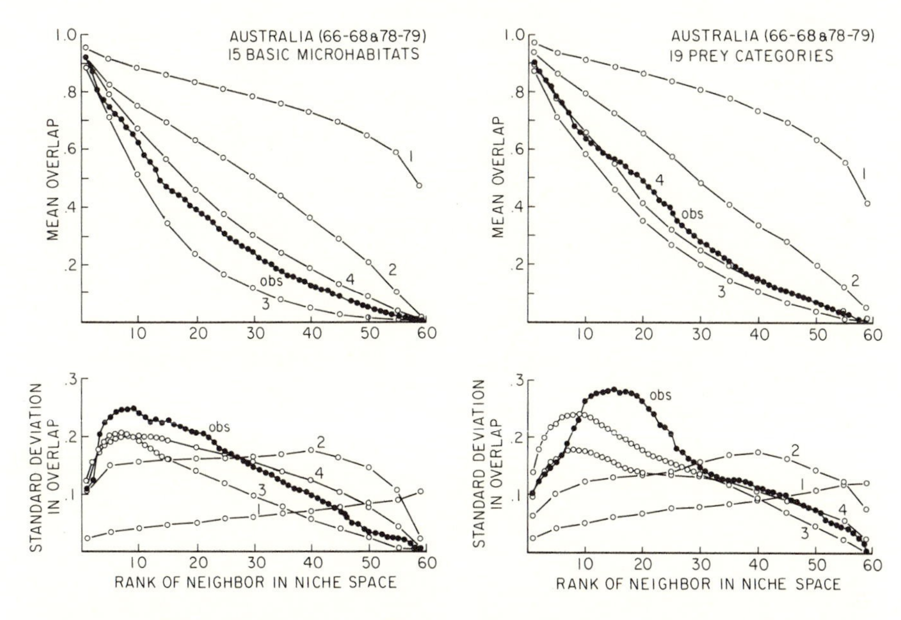


FIGURE 9.5. Plots like those shown in Figure 9.1, but for Australian lizards (microhabitat resources).

FIGURE 9.6. Plots like those shown in Figure 9.1, but for food resources of Australian desert lizards.

allowing a total of ninety alien-versus-resident comparisons. Resource utilization spectra of all other resident species are left exactly as observed. If a given resource state is missing from one community, it is excluded from the analysis (i.e., similarities are calculated using only resources that are actually present in both communities). As one possible measure of the "goodness of fit" among species, assume that the Lotka-Volterra competition equations adequately describe the system. Consumer species are thus assumed to have reached dynamic equilibria with one another, and observed relative abundances are proportional to equilibrium abundances. Moreover, observed dietary similarities are assumed to approximate competition coefficients. The diffuse competition load on a given target species is estimated by the summation (over all other species) of the products of the alphas times equilibrium densities. Each resident species' observed equilibrium population density is then compared to the theoretical population density that a transplanted alien of the same species would achieve if introduced in its place into its community. By this criterion, residents of most species definitely tend to achieve higher densities than do aliens (Tables 9.2, 9.3, and 9.4). Among all species over all study areas, residents outperformed aliens in 1,871 out of 3,014, or 62%, of such "experiments." This trend is much more pronounced when expanded food-resource matrices are used: for the Kalahari, when 46 prey categories (including termite castes) are recognized instead of 20, residents outperform aliens in 810 out of 1,056 cases (a full 76.7%). A comparable analysis for the two Australian study areas visited in 1978–79 (the L-area and Redsands), based on some 300 different prey resource states, was also undertaken: only 26 species were present on both sites, but residents outperformed aliens in 39 of the 52 possible introductions (75% of the time). These results strongly suggest that compensatory interactions are occurring among naturally coexisting species.

TABLE 9.2. Number of times that "residents" outperform "aliens," by species in North America (see text).

| SPECIES | NUMBER | TOTAL | PERCENTAGE |
|---|---|---|---|
| *Cnemidophorus tigris* | 61 | 90 | 67.8 |
| *Uta stansburiana* | 56 | 90 | 62.2 |
| *Crotaphytus wislizeni* | 55 | 90 | 61.1 |
| *Phrynosoma platyrhinos* | 62 | 90 | 68.9 |
| *Callisaurus draconoides* | 48 | 72 | 66.7 |
| *Coleonyx variegatus* | 25 | 42 | 59.5 |
| *Dipsosaurus dorsalis* | 7 | 12 | 58.3 |
| *Sceloporus magister* | 4 | 6 | 66.7 |
| *Urosaurus graciosus* | 1 | 6 | 16.7 |
| Overall totals | 319 | 498 | 64.1 |

TABLE 9.3. Number of times that "residents" outperform "aliens," by species in the Kalahari (see text).

| SPECIES | NUMBER | TOTAL | PERCENTAGE |
|---|---|---|---|
| *Agama hispida* | 64 | 90 | 71.1 |
| *Eremias lineo-ocellata* | 63 | 90 | 70.0 |
| *Eremias lugubris* | 37.5 (tie) | 72 | 52.1 |
| *Eremias namaquensis* | 22 | 42 | 52.4 |
| *Mabuya occidentalis* | 56 | 90 | 62.2 |
| *Mabuya striata* | 43 | 56 | 76.8 |
| *Meroles suborbitalis* | 45 | 72 | 62.5 |
| *Chondrodactylus angulifer* | 67 | 90 | 74.4 |
| *Pachydactylus capensis* | 49 | 72 | 68.1 |
| *Ptenopus garrulus* | 53 | 90 | 58.9 |
| *Typhlosaurus lineatus* | 47.5 (tie) | 72 | 66.0 |
| *Colopus wahlbergi* | 39.5 (ties) | 72 | 54.9 |
| *Mabuya variegata* | 26 | 42 | 61.9 |
| *Mabuya spilogaster* | 4 | 6 | 66.7 |
| *Pachydactylus rugosus* | 22.5 (tie) | 42 | 53.6 |
| *Ichnotropis squamulosa* | 1 | 2 | 50.0 |
| *Typhlosaurus gariepensis* | 20.5 (tie) | 30 | 68.3 |
| *Nucras tessellata* | 4 | 6 | 66.7 |
| *Pachydactylus bibroni* | 12 | 20 | 60.0 |
| Overall totals | 677 | 1056 | 64.1 |

TABLE 9.4. Number of times that "residents" outperform "aliens," by species in Australia (see text).

| SPECIES | NUMBER | TOTAL | PERCENTAGE |
|---|---|---|---|
| *Ctenophorus inermis* | 61 | 90 | 67.8 |
| *Pogona minor* | 32 | 56 | 57.1 |
| *Ctenophorus scutulatus* | 3 | 6 | 50.0 |
| *Moloch horridus* | 37 | 56 | 66.1 |
| *Varanus eremius* | 43.5 (tie) | 72 | 60.4 |
| *Varanus gouldi* | 57 | 90 | 63.3 |
| *Varanus tristis* | 35 | 56 | 62.5 |
| *Cryptoblepharus plagiocephalus* | 8 | 12 | 66.7 |
| *Morethia butleri* | 11 | 20 | 55.0 |
| *Menetia greyi* | 11 | 20 | 55.0 |
| *Lerista muelleri* | 1 | 2 | 50.0 |
| *Egernia inornata* | 24 | 56 | 42.9 |
| *Ctenotus grandis* | 38 | 56 | 67.9 |
| *Ctenotus leonhardii* | 4 | 6 | 66.7 |
| *Ctenotus pantherinus* | 42 | 72 | 58.3 |
| *Ctenotus schomburgkii* | 20 | 42 | 47.6 |
| *Delma fraseri* | 3 | 6 | 50.0 |
| *Lialis burtonis* | 6 | 12 | 50.0 |
| *Diplodactylus elderi* | 11 | 20 | 55.0 |
| *Diplodactylus strophurus* | 13 | 20 | 65.0 |
| *Gehyra variegata* | 43 | 72 | 59.7 |
| *Heteronotia binoei* | 36.5 (tie) | 56 | 65.2 |
| *Rhynchoedura ornata* | 59.5 (ties) | 90 | 66.1 |
| *Ctenophorus isolepis* | 41 | 72 | 56.9 |
| *Diporiphora winneckei* | 1 | 2 | 50.0 |
| *Gemmatophora longirostris* | 4 | 6 | 66.7 |
| *Varanus brevicauda* | 1 | 2 | 50.0 |
| *Egernia striata* | 32 | 56 | |
| *Lerista bipes* | 22.5 (tie) | 30 | 75.0 |
| *Ctenotus calurus* | 29.5 (tie) | 56 | 52.7 |
| *Ctenotus dux* | 4 | 6 | 66.7 |
| *Ctenotus helenae* | 26.5 | 56 | 47.3 |
| *Pygopus nigriceps* | 6 | 12 | 50.0 |
| *Diplodactylus conspicillatus* | 37 | 56 | 66.1 |
| *Diplodactylus damaeus* | 9 | 12 | 75.0 |
| *Nephrurus laevissimus* | 3 | 6 | 50.0 |
| *Ctenophorus clayi* | 1 | 2 | 50.0 |
| *Ctenophorus fordi* | 1 | 2 | 50.0 |
| *Lerista desertorum* | 1 | 2 | 50.0 |
| *Ctenotus brooksi* | 2 | 2 | 100.0 |

TABLE 9.4. *(continued)*

| SPECIES | NUMBER | TOTAL | PERCENTAGE |
|---|---|---|---|
| *Ctenotus colletti* | 1 | 2 | 50.0 |
| *Ctenotus piankai* | 18 | 30 | 60.0 |
| *Ctenotus quattuordecimlineatus* | 19 | 30 | 63.3 |
| *Ctenotus ariadnae* | 3 | 6 | 50.0 |
| *Diplodactylus ciliaris* | 4 | 6 | 66.7 |
| *Eremiascincus richardsoni* | 1.5 (tie) | 2 | 75.0 |
| *Omolepida branchialis* | 1.5 (tie) | 2 | 75.0 |
| *Nephrurus levis* | 6 | 12 | 50.0 |
| *Nephrurus vertebralis* | 1 | 2 | 50.0 |
| Overall totals | 876 | 1460 | 60.0 |

# 10 Dynamics: An Area Revisited

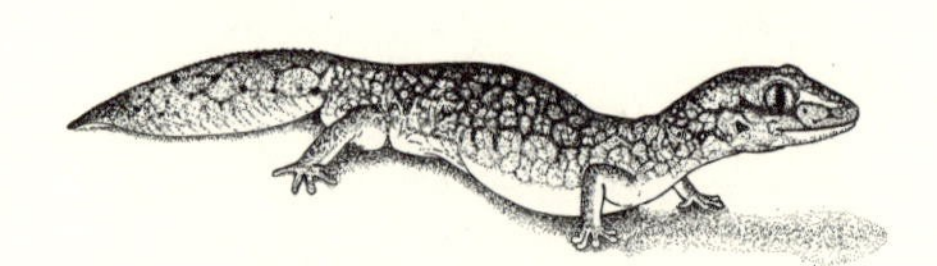

One wonders how stable these lizard assemblages might be. To try to assess the replicability of earlier results, I returned to the L-area in Australia more than ten years after first studying it in 1966–68. I wanted to determine whether resource utilization patterns were stable. I also hoped to characterize better the resources used by apparently "rare" species.

Little change was evident on the L-area in 1978–79. Data on resource utilization were collected and compared with other information gathered a full decade before. Considerable species-specificity and substantial fidelity in the use of food types and microhabitats is evident among many of these Australian desert lizards. For example, some species are termite specialists, whereas others virtually never touch termites; still other species eat ants to the exclusion of other prey. Diet and microhabitat utilization patterns of most species, even those of generalists such as the gecko *Gehyra variegata*, proved to be fairly consistent in time. Tables 10.1 and 10.2 summarize results obtained for eight relatively abundant species on the L-area in 1966–68 and again in 1978–79. Overall estimates based upon all the specimens of each species are even more conservative, varying relatively little between 1966–68 and 1978–79. Also, in site-to-site comparisons, each species is typically its own closest neighbor in niche space. Some shifts in resource utilization were also evident in certain species on the L-area, however (see also below).

On the L-area, 530 lizard specimens representing 27 species were collected in 1966–68. Tracks of the wary large monitor *Varanus gouldi* were regularly noted, although no specimens of these elusive lizards were sighted. The spinifex gecko *Diplodactylus elderi* was listed as "highly expected on the basis of geographic range, habitat, autecology, and microhabitat" (Pianka 1969a). During the 1978–79 expedition, some 1565 new lizard specimens representing 32 species were captured on the L-area. Only one species that was collected in 1966–68 was not encountered on the second trip (*Egernia kintorei*, a very uncommon large nocturnal skink); five new species were recorded, including *Varanus gouldi* and *Diplodactylus elderi* (both highly expected, as mentioned above). Tracks of the enormous *Varanus giganteus* were seen, but these exceedingly intelligent lizards always evaded sighting and easily eluded capture. Even though species richness went up between visits from 27 to 32 species, lizard species diversity actually decreased from 9.9 to 6.2, largely due to increases in the apparent abundance of a few species.

Relative abundances of various species on the L-area, as reflected in the numbers actually collected, did not remain constant but fluctuated fairly substantially (Table 10.3). Relative abundance decreased in 18 species and increased in 13 species. Abundances of some species shifted upward or downward by factors of 3 or more, but most changed less. Regardless of the direction (increase versus decrease), the average magnitude of change in relative abundance is 2.69 (st.dev. = 1.64, $N = 26$). Apparent changes in the abundances of the very uncommon species could easily be artifacts and probably should not be taken too seriously. While these differences are doubtless attributable, in part, to capricious collecting events as well as real differences between collectors, I nevertheless gained the distinct impression that at least two species had markedly increased in abundance (*Ctenophorus isolepis* and *Ctenotus calurus*), whereas several other species seemed to have declined drastically (*Ctenotus grandis, Ctenotus helenae, Gehyra variegata* and *Rhynchoedura ornata*). In the light of these apparent changes in relative abundance, closer scrutiny of resource utilization patterns among species is instructive (see Tables 10.1 and 10.2 and below).

Foods available to these lizards appear to have undergone some change between 1966–68 and 1978–79, at least as reflected in what the animals actually ate (Table 10.4). An average lizard stomach contained less food in 1978–79 (mean = .31 cc) than in 1966–68 (mean = .47 cc), suggesting an overall decrease in food availability. Dietary changes between the two visits may be deemed as relatively conservative because percentage compositions of prey categories changed by less than an order of magnitude. Nonetheless, grasshoppers/crickets and insect larvae dropped precipitously, whereas ants and vertebrates showed strong increases (the apparent change in vertebrate foods is an artifact to the extent that it reflects my own heightened effort to collect *Varanus*). Note that the diversity of foods eaten by all lizards

TABLE 10.1. Diets of 8 selected species (representing three families) of lizards on the L-area in 1966–68 and 1978–79 (in percentages, by volume).

| Prey Category | *Ctenophorus isolepis* | | *Moloch horridus* | | *Cryptoblepharus plagiocephalus* | |
|---|---|---|---|---|---|---|
| | 1966–68 | 1978–79 | 1966–68 | 1978–79 | 1966–68 | 1978–79 |
| Centipedes | | 0.2 | | | | |
| Spiders | 0.4 | 0.8 | | | 12.7 | 22.7 |
| Ants | 24.8 | 49.2 | 100.0 | 100.0 | | 9.1 |
| Wasps | 4.6 | 2.3 | | | | 1.2 |
| Grasshoppers/ crickets | 3.5 | 9.8 | | | 17.5 | 6.0 |
| Roaches | | 1.1 | | | | 9.1 |
| Mantids | | 0.2 | | | 7.8 | 1.2 |
| Beetles | 3.9 | 3.7 | | | 4.3 | 16.9 |
| Termites | 33.4 | 23.9 | | | 15.8 | 10.6 |
| Hemipterans | 5.0 | 2.4 | | | 16.2 | 6.0 |
| Diptera | | 3.0 | | | 1.0 | 7.9 |
| Lepidoptera | 1.0 | | | | 3.9 | |
| Insect Larvae | 20.5 | 1.9 | | | 6.8 | |
| Miscellaneous unidentified insects | 3.1 | 0.6 | | | 14.0 | 5.4 |
| Vertebrates | | | | | | 3.3 |
| Plant material | | 0.7 | | | | |
| Total volume of prey, cc. | 13.48 | 97.66 | 5.70 | 3.86 | 0.51 | 1.66 |

| *Egernia striata* | | *Ctenotus grandis* | | *Ctenotus helenae* | | *Gehyra variegata* | | *Rhynchoedura ornata* | |
|---|---|---|---|---|---|---|---|---|---|
| 1966–68 | 1978–79 | 1966–68 | 1978–79 | 1966–68 | 1978–79 | 1966–68 | 1978–79 | 1966–68 | 1978–79 |
| | 1.5 | | 6.6 | | 0.5 | | | | |
| 0.1 | 0.8 | 1.3 | 0.2 | 0.9 | 2.3 | 6.6 | 3.1 | | |
| 3.8 | 8.1 | 2.8 | 0.7 | 1.0 | 0.7 | 1.2 | 0.4 | 0.4 | 0.5 |
| 0.1 | 0.4 | | | 1.1 | 0.4 | 1.8 | 0.2 | | |
| | | | | | | | | | |
| 0.2 | 0.8 | 0.1 | | 5.3 | 5.6 | 9.5 | 7.3 | | |
| 0.3 | 3.7 | 1.5 | | 10.1 | 6.7 | 4.1 | 21.2 | | |
| | | | | | | 0.5 | 0.7 | | |
| 2.1 | 6.3 | 0.5 | 2.2 | 1.5 | 4.3 | 7.3 | 18.5 | | |
| 86.2 | 70.3 | 72.9 | 89.2 | 69.6 | 73.7 | 49.4 | 15.1 | 99.4 | 96.7 |
| 0.1 | 0.2 | | 0.05 | | 0.2 | 6.9 | 9.7 | | |
| | | | | | | | 0.2 | | |
| 0.3 | | 0.8 | | 0.6 | 2.6 | 1.7 | 6.5 | | |
| 2.8 | 0.05 | 7.0 | | 7.1 | 0.2 | 5.5 | 3.9 | | |
| | | | | | | | | | |
| 0.7 | 3.1 | 1.0 | 0.8 | 0.7 | | 2.7 | 4.8 | | 2.4 |
| 3.1 | 1.5 | 6.5 | | | 2.7 | 0.4 | 8.0 | | |
| 0.2 | 3.1 | 5.5 | | 2.1 | 0.2 | | | 0.1 | 0.5 |
| | | | | | | | | | |
| 37.93 | 53.76 | 50.64 | 61.54 | 40.69 | 17.10 | 12.26 | 21.85 | 2.30 | 2.10 |

TABLE 10.2. Percentage utilization of various microhabitats among 8 selected lizard species (representing three families) on the L-area in 1966–68 and 1978–79.

| MICROHABITAT CATEGORY | *Ctenophorus isolepis* | | *Moloch horridus* | | *Cryptoblepharus plagiocephalus* | |
|---|---|---|---|---|---|---|
| | 1966–68 | 1978–79 | 1966–68 | 1978–79 | 1966–68 | 1978–79 |
| Open sun | 66.0 | 32.6 | 71.4 | 42.9 | 3.7 | |
| Grass sun | 20.8 | 55.2 | | 21.4 | 3.7 | |
| Bush sun | | 0.4 | | | | |
| Tree sun | | 0.5 | | | 5.6 | 1.9 |
| Other sun | | | | | | 10.4 |
| Open shade | | 0.6 | | | 3.7 | |
| Grass shade | 13.2 | 9.6 | 28.6 | 7.1 | | |
| Bush shade | | 0.6 | | | | |
| Tree shade | | 0.4 | | 28.6 | 1.9 | |
| Other shade | | | | | | 0.9 |
| Low sun | | | | | 25.0 | 36.8 |
| Low shade | | 0.2 | | | 13.9 | 6.6 |
| High sun | | | | | 30.6 | 34.0 |
| High shade | | | | | 12.0 | 9.4 |
| Total number of lizards | 53 | 513 | 7 | 7 | 27 | 53 |

| *Egernia striata* | | *Ctenotus grandis* | | *Ctenotus helenae* | | *Gehyra variegata* | | *Rhynchoedura ornata* | |
|---|---|---|---|---|---|---|---|---|---|
| 1966–68 | 1978–79 | 1966–68 | 1978–79 | 1966–68 | 1978–79 | 1966–68 | 1978–79 | 1966–68 | 1978–79 |
| 2.6 | 28.6 | 11.1 | 21.9 | 6.4 | 7.1 | | | | |
| 5.3 | 9.5 | 8.9 | 12.5 | 6.4 | 14.3 | | | | |
| | 4.8 | | | | | | | | |
| | | | | 2.1 | | | | | |
| 9.2 | 21.4 | | | | | | | | |
| 27.6 | | | | | | 3.5 | 1.5 | 86.0 | 93.5 |
| 26.3 | 9.5 | 80.0 | 65.6 | 75.5 | 71.4 | | 3.0 | 2.0 | 3.2 |
| 5.3 | 4.8 | | | | | 0.9 | 1.5 | | |
| | | | | 9.6 | 3.6 | 3.9 | 3.7 | | 3.2 |
| 23.7 | 21.4 | | | | | 0.9 | 5.7 | 12.0 | |
| | | | | | | | | | |
| | | | | | | 39.0 | 39.5 | | |
| | | | | | 3.6 | | | | |
| | | | | | | 51.8 | 45.0 | | |
| | | | | | | | | | |
| 37 | 21 | 45 | 32 | 47 | 28 | 114 | 200 | 50 | 62 |

TABLE 10.3. Number of lizards collected on the L-area in 1966–68 and 1978–79, with their relative abundances (in percentages of total numbers of lizards of all species).

| SPECIES | 1966–68 | | 1978–79 | | CHANGE IN RELATIVE ABUNDANCE |
|---|---|---|---|---|---|
| | *Number* | % | *Number* | % | |
| *Ctenophorus inermis* | 9 | 1.70 | 5 | 0.32 | −5.30 |
| *Ctenophorus isolepis* | 55 | 10.38 | 530 | 33.87 | +3.30 |
| *Moloch horridus* | 9 | 1.70 | 9 | 0.58 | −2.90 |
| *Pogona minor* | 8 | 1.51 | 12 | 0.77 | −1.96 |
| *Varanus eremius* | 3 | 0.57 | 13 | 0.83 | +1.46 |
| *Varanus gouldi* | tr | — | 4 | 0.26 | + |
| *Varanus tristis* | 6 | 1.13 | 18 | 1.15 | +1.02 |
| *Ctenotus ariadnae* | 5 | 0.94 | 11 | 0.70 | −1.34 |
| *Ctenotus calurus* | 11 | 2.08 | 147 | 9.39 | +4.50 |
| *Ctenotus grandis* | 45 | 8.49 | 39 | 2.49 | −3.40 |
| *Ctenotus helenae* | 53 | 10.00 | 31 | 1.98 | −5.05 |
| *Ctenotus pantherinus* | 8 | 1.51 | 21 | 1.34 | −1.13 |
| *Ctenotus piankai* | 2 | 0.38 | 3 | 0.19 | −2.00 |
| *Ctenotus quattuordecimlineatus* | 46 | 8.68 | 183 | 11.69 | +1.35 |
| *Ctenotus schomburgkii* | 7 | 1.32 | 40 | 2.56 | +1.94 |
| *Cryptoblepharus plagiocephalus* | 27 | 5.09 | 53 | 3.39 | −1.50 |
| *Egernia kintorei* | 1 | 0.19 | 0 | — | — |
| *Egernia inornata* | 2 | 0.38 | 2 | 0.13 | −2.90 |
| *Egernia striata* | 37 | 6.98 | 68 | 4.35 | −1.60 |
| *Lerista bipes* | 6 | 1.13 | 53 | 3.39 | +3.00 |
| *Lerista muelleri* | 0 | — | 6 | 0.38 | + |
| *Menetia greyii* | 3 | 0.57 | 4 | 0.26 | −2.20 |
| *Morethia butleri* | 0 | — | 1 | 0.06 | + |
| *Delma fraseri* | 0 | — | 4 | 0.26 | + |
| *Lialis burtonis* | 2 | 0.38 | 2 | 0.13 | −1.10 |
| *Pygopus nigriceps* | 8 | 1.51 | 3 | 0.19 | −7.95 |
| *Diplodactylus conspicillatus* | 6 | 1.13 | 27 | 1.73 | +1.50 |
| *Diplodactylus elderi* | 0 | — | 4 | 0.26 | + |
| *Diplodactylus stenodactylus* | 0 | — | 1(2)* | 0.06 | + |
| *Gehyra variegata* | 114 | 21.51 | 202 | 12.91 | −1.67 |
| *Heteronotia binoei* | 1 | 0.19 | 1 | 0.06 | −3.17 |
| *Nephrurus levis* | 6 | 1.13 | 4 | 0.26 | −4.35 |
| *Rhynchoedura ornata* | 50 | 9.43 | 64 | 4.09 | −2.31 |
| Total number | 530 + tracks | | 1565 | | |

* One was collected, another was found inside the stomach of another lizard.

TABLE 10.4. Percentage composition of various prey categories in the overall diet of the entire L-area saurofauna in 1966–68 and 1978–79.

| PREY CATEGORY | 1966–68 | 1978–79 | CHANGE IN RELATIVE IMPORTANCE |
|---|---|---|---|
| Centipedes | 0.3 | 1.2 | +4.0 |
| Spiders | 1.2 | 0.7 | −1.7 |
| Scorpions | 1.5 | 1.7 | +1.1 |
| Ants | 5.5 | 12.4 | +2.3 |
| Wasps | 0.6 | 0.6 | 1.0 |
| Grasshoppers and crickets | 13.3 | 7.5 | −1.8 |
| Roaches | 2.5 | 2.9 | +1.2 |
| Mantids and Phasmids | 0.2 | 0.4 | +2.0 |
| Neuropterans | — | 0.1 | (+) |
| Beetles | 3.1 | 4.4 | +1.4 |
| Termites | 50.4 | 41.8 | −1.2 |
| Hemipterans | 1.2 | 1.6 | +1.1 |
| Diptera | 0.01 | 0.7 | +70.0 |
| Lepidoptera | 1.2 | 0.4 | −3.0 |
| Insect eggs and pupae | 0.2 | 0.1 | −2.0 |
| All insect larvae | 5.7 | 1.0 | −5.7 |
| Miscellaneous arthropods | 1.8 | 3.0 | +1.7 |
| Vertebrates | 5.2 | 18.3 | +3.5 |
| Plant materials | 6.0 | 1.4 | −4.3 |
| Total volume of food, cc. | 249.1 | 477.5 | |
| Diversity of foods eaten by all lizards | 3.48 | 4.28 | |

increased slightly, partially due to the decrease in the importance of termites (the L-area is unusual among my Australian study areas in its very high values for termite consumption).

Although there is no reason to suspect that availabilities of microhabitats should have altered appreciably over the decade between my two visits, Table 10.5 summarizes data on microhabitat utilization of the entire L-area saurofauna in 1966–68 compared to 1978–79 (comparable to Table 10.4 for foods eaten). The fraction of lizards first sighted in the sunshine at the edge of porcupine grass tussocks increased 4-fold, whereas those observed in grass shade decreased. Overall diversity of microhabitats used by all individuals of all lizard species declined.

TABLE 10.5. Percentage utilization of various microhabitats by the entire L-area lizard fauna in 1966–68 and 1978–79.

| MICROHABITAT CATEGORY | 1966–68 | 1978–79 | CHANGE IN RELATIVE IMPORTANCE |
|---|---|---|---|
| Subterranean | 1.2 | 1.9 | +1.6 |
| Open sun | 14.4 | 18.6 | +1.3 |
| Grass sun | 6.8 | 30.8 | +4.5 |
| Bush sun | 0.1 | 0.2 | +2.0 |
| Tree sun | 1.0 | 0.4 | −2.5 |
| Other sun | 0.8 | 0.8 | 1.0 |
| Open shade | 14.9 | 5.9 | −2.5 |
| Grass shade | 28.5 | 20.5 | −1.4 |
| Bush shade | 0.7 | 0.6 | −1.2 |
| Tree shade | 2.3 | 1.6 | −1.4 |
| Other shade | 3.4 | 1.6 | −2.1 |
| Low sun | 1.5 | 1.7 | +1.1 |
| Low shade | 9.4 | 6.0 | −1.6 |
| High sun | 2.1 | 1.8 | −1.2 |
| High shade | 13.0 | 7.5 | −1.7 |
| Total number of lizards | 516 | 1423 | |
| Diversity of microhabitats used by all lizards | 6.36 | 5.37 | |

A simplistic first hypothesis might be that abundances fluctuate directly with prey availability. If so, a doubling of the availability of ants would be expected to lead to a doubling of the density of myrmecophagous species. Likewise, abundances of termite specialists would be expected to "track" termite availabilities. Provided one can assume that values in Table 10.4 reflect real changes in availabilities, this hypothesis is easily tested and rejected: *Moloch horridus*, an obligate ant specialist, decreased by a factor of three even though ants increased more than 2-fold. Also, although termites decreased slightly (from 50% to 42%) in the overall diet of all lizards (Table 10.4), the relative abundance of an obligate termite specialist *Diplodactylus conspicillatus* increased by 50% (another termite specialist, *Rhynchoedura ornata*, fluctuated in the opposite direction, decreasing to less than half its former abundance). A related observation of interest can be made on *Ctenotus calurus*, the species that exhibited the most dramatic increase (450%): this tiny blue-tailed skink almost doubled its consumption of termites from 1966–68 to 1978–79 (from 44.3% to 81.2%), in spite of the fact that termites decreased in the overall diet of all lizards. The fraction of insect larvae in its diet fell from 51.3% to only 2% (percentage representation of larvae in the overall diet of all lizards fell from 5.7% to 1.0%).

Yet another interesting, although unfortunately uncommon, species is the flap-footed legless lizard, *Pygopus nigriceps*, a nocturnal denizen of the open spaces with an unusually high consumption of scorpions (the diet of 16 individuals consisted of 34% scorpions by volume). In the overall diet of all lizards, the importance of scorpions was trivial and did not change appreciably (only 1.5% to 1.7%; Table 10.4). Nonetheless, *Pygopus* declined drastically in relative abundance from 1.5% to a mere 0.2%.

Dietary and microhabitat niche breadths, and changes therein, are summarized in Table 10.6. The average magnitude of observed change in niche breadths (irrespective of sign) among all species was similar for foods (1.51) and microhabitats (1.46). Average food niche breadths declined slightly (2.55 to 2.13), whereas average microhabitat niche breadth remained approximately constant (2.30 versus 2.32). Variances in niche breadths among species were lower in microhabitat than in diet (Table 10.6).

A prediction of optimal foraging theory is that diets should tend to contract when foods are abundant but should expand when foods are scarce (MacArthur and Pianka 1966). One might also expect relative abundances of consumers to vary directly with prey abundance. Of the seven species that increased in relative abundance, food niche breadths decreased as predicted in four, increased in two, and stayed constant in one (the termite specialist *Diplodactylus conspicillatus*). Two species changed little in relative abundance and their food niche breadths remained fairly constant. Of the 15 species that declined in abundance, diets expanded (as expected) in only two, contracted in six, and changed little among seven others (the latter include both food specialists and food generalists).

These data may also be exploited to test the hypotheses that abundances of ecologically similar species fluctuate either (1) in phase with one another, or (2) out of phase with one another, more so as compared with ecologically more dissimilar species. The first hypothesis emerges from a noncompetition argument asserting only that species track resources, whereas the second might suggest coactions between species of the mutually detrimental class such as might arise from interspecific competition. Both hypotheses could easily be fatal oversimplifications if the majority of species experience diffuse competition and positive as well as negative indirect effects on one another (see also below).

In an effort to perform such a test, I computed the direction and

TABLE 10.6. Dietary and microhabitat niche breadths, calculated from data like those in Tables 10.1 and 10.2, using Simpson's (1949) index of diversity.

| | DIET | | | MICROHABITAT | | |
|---|---|---|---|---|---|---|
| SPECIES | 1966–68 | 1978–79 | *Change* | 1966–68 | 1978–79 | *Change* |
| *Ctenophorus inermis* | 4.33 | 2.89 | −1.5 | 1.80 | 5.00 | +2.8 |
| *Ctenophorus isolepis* | 4.48 | 3.19 | −1.4 | 2.01 | 2.38 | +1.2 |
| *Moloch horridus* | 1.00 | 1.00 | 1.0 | 1.69 | 3.16 | +1.9 |
| *Pogona minor* | 3.50 | 3.51 | 1.0 | 3.84 | 3.90 | +1.02 |
| *Varanus eremius* | 1.41 | 2.14 | +1.5 | 4.03 | 2.89 | −1.4 |
| *Varanus gouldi* | 2.77 | 2.44 | −1.1 | 1.55 | 3.60 | +2.3 |
| *Varanus tristis* | 1.34 | 1.46 | +1.1 | 1.92 | 2.13 | +1.1 |
| *Ctenotus ariadnae* | 1.42 | 1.33 | −1.07 | 2.94 | 1.86 | −1.6 |
| *Ctenotus calurus* | 2.17 | 1.30 | −1.67 | 1.97 | 2.58 | +1.3 |
| *Ctenotus grandis* | 1.85 | 1.25 | −1.47 | 1.52 | 2.02 | +1.3 |
| *Ctenotus helenae* | 1.99 | 1.80 | −1.11 | 1.70 | 1.86 | +1.1 |
| *Ctenotus pantherinus* | 1.30 | 1.04 | −1.25 | 1.47 | 1.65 | +1.1 |
| *Ctenotus piankai* | 3.57 | 1.47 | −2.43 | 2.00 | 1.00 | −2.0 |
| *Ctenotus quattuordecimlineatus* | 5.51 | 1.55 | −3.55 | 1.49 | 2.10 | +1.4 |
| *Ctenotus schomburgkii* | 1.01 | 1.30 | +1.29 | 3.38 | 3.36 | −1.01 |
| *Cryptoblepharus plagiocephalus* | 7.59 | 7.97 | +1.05 | 5.07 | 3.63 | −1.4 |
| *Egernia kintorei* | 1.44 | — | — | 1.00 | — | — |
| *Egernia inornata* | 1.17 | 1.00 | −2.3 | 5.47 | 2.67 | −2.1 |
| *Egernia striata* | 1.34 | 1.97 | +1.5 | 4.63 | 5.10 | +1.1 |
| *Lerista bipes* | 3.83 | 1.90 | −2.0 | 1.00 | 1.28 | +1.3 |
| *Lerista muelleri* | — | 3.58 | — | — | 1.39 | — |
| *Menetia greyii* | 2.00 | 1.29 | −1.55 | 2.57 | 1.60 | −1.6 |
| *Morethia butleri* | — | 1.00 | — | — | 2.00 | — |
| *Delma fraseri* | — | 1.92 | — | — | 1.00 | — |
| *Lialis burtonis* | 1.13 | — | — | 1.00 | 2.00 | +2.0 |
| *Pygopus nigriceps* | 1.49 | 1.47 | −1.01 | 1.28 | 2.00 | −1.6 |
| *Diplodactylus conspicillatus* | 1.00 | 1.02 | +1.02 | 1.80 | 2.31 | +1.3 |
| *Diplodactylus elderi* | — | 2.47 | — | — | 1.00 | — |
| *Diplodactylus stenodactylus* | — | 1.00 | — | — | 1.00 | — |
| *Gehyra variegata* | 3.64 | 7.57 | +2.1 | 2.36 | 2.74 | +1.2 |
| *Heteronotia binoei* | 1.80 | 1.00 | −1.8 | 1.00 | 1.00 | 1.0 |
| *Nephrurus levis* | 5.30 | — | — | 2.67 | 3.00 | +1.1 |
| *Rhynchoedura ornata* | 1.01 | 1.07 | +1.06 | 1.33 | 1.14 | −1.2 |
| Mean | 2.55 | 2.13 | 1.51* | 2.30 | 2.32 | 1.46* |
| Standard deviation | 1.68 | 1.71 | 0.60* | 1.27 | 1.10 | 0.45* |

* Based on absolute values.

magnitude of changes in relative abundances among all possible pairs of species. Using lumped data from both visits, overall ecological overlap was estimated as the product of dietary overlap times microhabitat overlap. The relative change in the abundances of each pair of species was expressed as the ratio of the change in each ($\Delta N_i / \Delta N_j$). No correlation emerged from comparison of this matrix of changes in abundance versus the above-mentioned matrix of overall ecological overlap, either among all 33 species ($r = -.011$) or using just 11 species for which sample sizes are more adequate ($r = -.001$). Nor do the elements in the matrix of changes in abundance correlate with dietary overlap ($r = +.006$). Finally, the correlation between the elements in the above matrix of changes in abundance versus elements in the inverse of the matrix of overall ecological overlap is also negligible ($r = +.005$). While these negative results are less than satisfying, at least they seem to support neither of the above hypotheses. Rather, they suggest either that stochasticity in this system is considerable or that abundances of each species vary more or less independently of those in other species. Alternatively, abundances of species might vary in response to resources and one another; *but* their interactions might be aligned along many niche dimensions, with some direct competitors yielding positive density effects (indirect "competitive mutualisms"), whereas others yield net negative density effects. Under such complex circumstances, overall total effects could appear to be mere "noise."

### *Fire Succession Cycle*

I returned briefly to the L-area once more in May-June of 1984 to discover that it had been partially burned (as have nearly all of my Australian study areas since 1968). I am now firmly convinced that a fire succession cycle occurs, which would repay long-term study. The time dimension of these dynamics is measured in decades, making them difficult to appreciate. As more or less regular agents of disturbance, fires probably contribute substantially to maintaining diversity in Australian desert-lizard systems. These fires are usually started by lightning, raging completely out of control for weeks on end across many square kilometers of desert. Fires vary considerably in intensity and extensiveness. *Eucalyptus* trees are fire resistant and frequently survive a hot ground fire carried by the exceedingly flammable *Triodia* grass tussocks. Moreover, fires typically reticulate, missing large tracts immediately adjacent to burns. Effects on lizards and lizard microhabitats are drastic yet heterogeneous in space. Some lizard species with open habitat requirements, such as *Ctenophorus inermis* and *Ctenotus calurus*, are able to reinvade burned areas rapidly, whereas other species, such as *Ctenotus helenae, Delma fraseri*, and/or *Omolepida branchialis*, which require large spinifex tussocks for microhabitats, cannot. However, such species may continue to exist in isolated pockets and patches of unburned habitat. Spinifex rejuvenates rapidly, probably from live roots as well as by seedling establishment. Newly burned areas are very open with lots of bare ground and tiny, well-spaced clumps of *Triodia*. Unburned patches, in contrast, are composed of large ancient tussocks, frequently close together with little open space between them. As time progresses, *Triodia* clumps grow and "close in," gradually becoming more and more vulnerable to carrying another fire. Throughout this process, lizard microhabitats (and associated food resources?) would change, too. The elements are clearly present for a dynamic and complex model of succession, incorporating disturbance probabilities, the resulting spatial heterogeneity of microhabitats, and species-specific habitat requirements and colonization abilities.

# 11 Anatomical Correlates of Ecology

Population biologists have long appreciated that an animal's morphology reflects its ecology (for a review, see Hespenheide 1973). Desert lizards are no exception. Species that spend a lot of time in the open away from cover tend to have longer hind legs, relative to their snout-vent length, than those that stay closer to safe retreats (Figure 11.1). Longer legs doubtlessly increase running speed and hence facilitate the use of open spaces. However, long-legged species move very clumsily through dense vegetation, demonstrating that there is actually a premium on shorter legs for species that exploit such closed-in microhabitats. Terrestrial species also tend to have longer hind legs than arboreal species in *Sceloporus* (Lundelius 1957) and in *Anolis* (Collette 1961). Ananjeva (1977) found differences in limb proportions among five sympatric lacertids (genus *Eremius*), which proved to be correlated with their adaptations for climbing, burrowing, and other movements. Digging species typically possess larger front feet and more powerful forelegs than species that do not do much digging with their front limbs. Fossorial species have reduced appendages or lack them altogether.

Numerous other morphological correlates of the use of space exist. Arboreal lizards are typically long-tailed and slender, with claws or toe lamellae well suited for climbing. Indeed, the number of toe lamellae, as well as their surface area, are thought to be intimately related to climbing ability (Collette 1961; Hecht 1952). Among nocturnal geckos, terrestrial species tend to have larger eyes than arboreal species (Werner 1969; Pianka and Pianka 1976).

Head proportions and jaw length are often reasonably accurate indicators of the size of a lizard's prey (Figures 11.2 and 11.3). Lizard dentition also reflects more subtle aspects of diet, such as the agility and hardness of prey (Hotton 1955).

An obvious hope is that such ecomorphological correlates will ultimately enable ecological predictions based on anatomical data (Ricklefs and Travis 1980). Another suggestion is that morphological measures can be estimated more objectively than ecological parameters. Still another idea is that morphometrics may represent average long-term responses to selection and hence reflect environmental conditions better than more direct measurements of the ever-changing immediate ecological milieu.

Van Valen (1965) postulated that morphologically variable species should often be ecological generalists by virtue of a high between-phenotype component of niche breadth (see also Roughgarden 1972). In contrast, species that are ecological specialists would be expected to show low morphological variability and little variance between individuals in resource utilization. Empirical support for this morphological variation-niche width hypothesis has been slow in coming; most attempts to test the hypothesis have relied on avian beak meas-

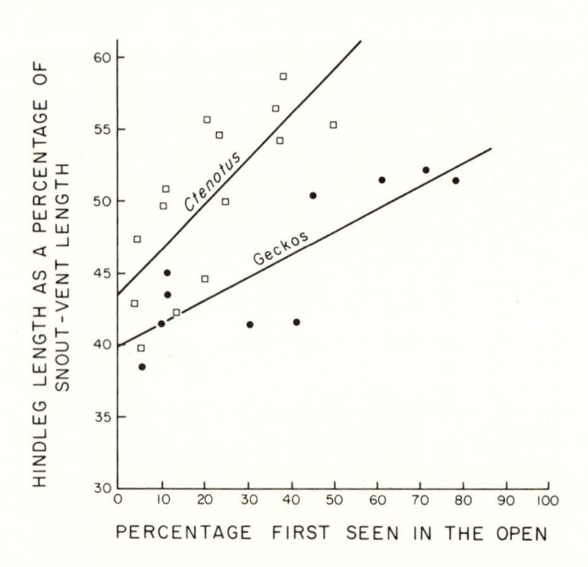


FIGURE 11.1. Hind-leg length expressed as a percentage of snout-vent length plotted against the percentage of lizards first sighted in the open spaces between plants among 14 species of Australian *Ctenotus* skinks (open squares) and 10 species of Australian geckos (closed circles).

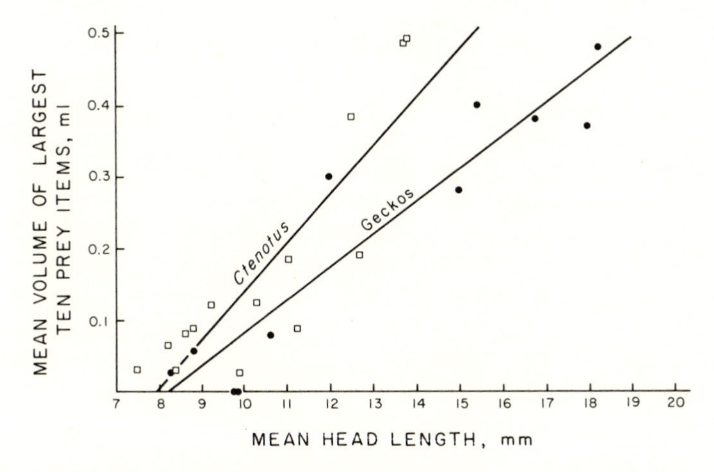


FIGURE 11.2. Mean volume of the ten largest prey items plotted against average head length for 14 species of *Ctenotus* skinks (open squares) and 10 species of Australian geckos (closed circles).

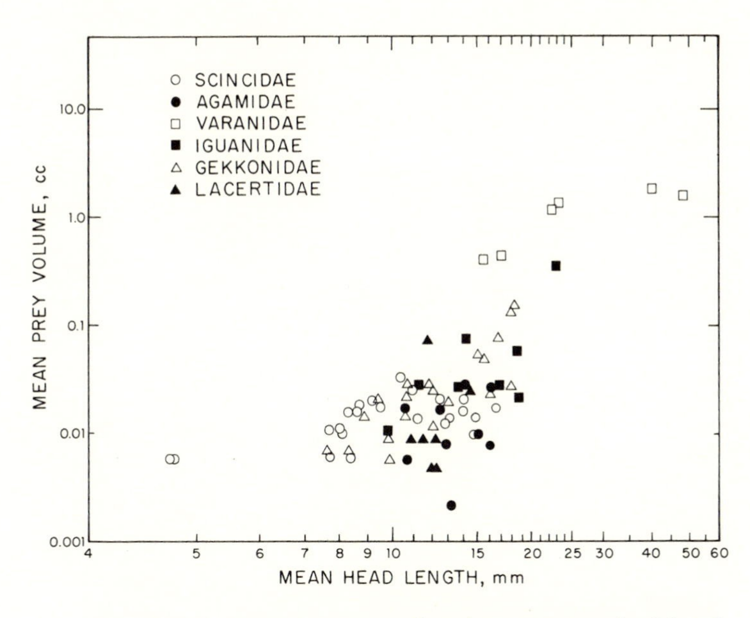


FIGURE 11.3. Average prey size plotted against mean head length among 74 species of desert lizards in six different families (log-log plot). The correlation with mean size of the ten largest prey items is even stronger.

urements (Grant 1967, 1971; Soule and Stewart 1970; Van Valen and Grant 1970; Willson 1969).

Size differences between closely related sympatric species have been implicated as being necessary for coexistence (Hutchinson 1959; MacArthur 1972; Schoener 1965, 1984), and even in the "assembly" of communities (Case et al. 1983), although there has been considerable dispute over the statistical validity of these patterns (Grant 1972; Horn and May 1977; Strong et al. 1979; Grant and Abbott 1980; Simberloff and Boecklen 1981).

Perhaps the most thorough analysis of such "Hutchinsonian" ratios is that provided for the world's bird-eating hawks by Schoener (1984), who computed size ratios among all possible pairs and triplets of the 47 species of accipiter hawks. Frequency distributions of expected size ratios were generated for all possible combinations of species, which were then compared with the much smaller number of existing accipiter assemblages. Schoener found a distinct paucity of low size ratios among real assemblages, strongly suggesting size assortment.

Although I certainly cannot provide anything even approaching a case as convincing as Schoener's, Table 11.1 reports somewhat comparable results of such an analysis of jaw lengths for the lizards considered here. There is no significant difference between all possible intracontinental pairs versus all possible intercontinental pairs. As

TABLE 11.1. Statistics on jaw lengths and Hutchinsonian ratios of jaw lengths (larger/smaller) for various lizard assemblages.

| | JAW LENGTH | | | HUTCHINSONIAN RATIO | | |
|---|---|---|---|---|---|---|
| ASSEMBLAGE | $\bar{x}$ | S.D. | N | $\bar{x}$ | S.D. | N |
| North America | 16.6 | 5.4 | 11 | 1.53 | 0.4 | 55 |
| Kalahari | 13.0 | 4.3 | 20 | 1.72 | 0.9 | 190 |
| Lacertids | 13.2 | 0.8 | 6 | 1.07 | 0.1 | 15 |
| *Mabuya* | 15.2 | 4.3 | 4 | 1.52 | 0.5 | 6 |
| Australia | 16.4 | 1.3 | 60 | 1.82 | 1.0 | 1770 |
| *Ctenotus* | 11.9 | 2.4 | 14 | 1.28 | 0.2 | 91 |
| *Diplodactylus* | 13.4 | 3.4 | 7 | 1.35 | 0.3 | 21 |
| All species (all continents) | 15.8 | 0.9 | 91 | 1.78 | 1.0 | 4095 |
| All intracontinental pairs | | | | 1.80 | 1.0 | 2015 |
| All intercontinental pairs | | | | 1.76 | 1.0 | 2080 |

NOTE: *Heloderma, Chameleo,* and *Varanus giganteus* are excluded.

might be expected, however, closely related groups do tend to exhibit lower ratios than those observed in entire saurofaunas.

Because anatomical parameters are usually much easier to estimate objectively than ecological ones, a variety of recent studies attempt to exploit such morphological correlates of ecology to make anatomical maps of ecological space and, in turn, to use these to analyze various aspects of community structure; these efforts deal with vertebrate taxa as divergent as bats (Findley 1973, 1976), birds (Karr and James 1975; Ricklefs and Travis 1980), fish (Gatz 1979a,b), and lizards[1] (Ricklefs et al. 1981). In this approach, each species is represented as a point in an *n*-dimensional hypervolume whose coordinates are the morphological variates. These may be standardized as desired or log transformed. Euclidean distances between species are calculated as measures of dissimilarity. Distances from the centroid of the hypervolume can be exploited to judge the overall size of morphological space. If desired, dimensionality can be reduced and orthogonality achieved by a multivariate procedure such as principal components analysis (note that Euclidean distances remain unchanged when axes are rotated). Spacing patterns between species, such as nearest-neighbor distances, and other aspects of their position in morphological space can then be examined. The assumption is usually made that the arrangement of species in morphological space accurately reflects their ecological relationships, although this assumption is not easily verifiable and has seldom been directly tested (but see Ricklefs and Travis 1980 and below).

For most individual lizards collected, ten morphometrics were measured: snout-vent and tail length; the length, width, and depth of the head; and the lengths of the jaw, forefoot, forearm, hind foot, and hind leg. Even though sexual dimorphisms do occur in some species, sexes are lumped for simplicity and ease of analysis here. These morphometrics proved to have strong positive correlations with one another over all 90-odd species (mean correlation coefficient = .75, st.dev. = .148); this high degree of correlation indicates substantial redundancy, calling for a multivariate analysis to reduce dimensionality (see below). Anatomical statistics for 94 species are summarized in Appendix G. Average morphometrics were used to represent each species as a point in a multidimensional morphospace. Each morphometric was given equal weight through standardization by subtracting the mean value for all 90-odd species and dividing by the standard deviation across all species: this standardization procedure

[1] This study, based on my own data on lizard anatomy, is revised and extended in the present chapter.

results in a mean score of 0.0 and a standard deviation of 1.0 for each variate.[2] Distances from this standardized hypervolume's centroid (representing the overall "average" lizard species) were calculated for each species and averaged for each continental saurofauna. Euclidean distances between all pairs of species were computed, and nearest-neighbor distances identified for each. Various morphometric statistics, computed for each continent separately as well as for all intracontinental plus intercontinental pairs of species, are summarized in Table 11.2. Anatomically, Kalahari lizard species are appreciably more similar than are the North American or Australian lizard species. Both the overall average and nearest neighbor Euclidean distances are smaller and less variable for Kalahari lizards than they are for the other two continental-desert lizard systems. Euclidean distances between species are most variable in Australia, probably partially due to the larger number of species there. The overall volume of mor-

TABLE 11.2. Morphometric statistics for various lizard assemblages.

| | NORTH AMERICA | KALAHARI | AUSTRALIA | INTRA-CONTINENTAL PAIRS | INTER-CONTINENTAL PAIRS | ALL PAIRS |
|---|---|---|---|---|---|---|
| Euclidean distances between species | | | | | | |
| Mean | 31.7 | 21.4 | 35.4 | 34.0 | 31.9 | 32.9 |
| St. dev. | 15.9 | 12.2 | 30.6 | 29.3 | 24.0 | 26.7 |
| *N* (pairs) | 55 | 190 | 1770 | 2015 | 2080 | 4095 |
| Nearest neighbor | | | | | | |
| Mean | 7.5 | 4.0 | 8.0 | — | — | 7.0 |
| St. dev. | 3.4 | 1.8 | 9.8 | — | — | 8.2 |
| *N* (species) | 11 | 20 | 60 | — | — | 91 |
| Centroid distance | | | | | | |
| Mean | 26.4 | 16.1 | 24.1 | — | — | 22.6 |
| St. dev. | 11.4 | 10.3 | 22.6 | — | — | 19.6 |
| *N* (species) | 11 | 20 | 60 | — | — | 91 |
| Correlation between morphometric distance and ecological overlap (product) | | | | | | |
| $r$ | −.121 | −.474 | −.205 | | −.201 | −.212 |

[2] In an earlier effort (Ricklefs et al. 1981), morphometrics were log-transformed, which required omission of aberrant legless skinks and pygopodids. These species are included in the present analysis.

phospace occupied by Kalahari lizards, as judged by distances from the centroid (Table 11.2), is more compact than that occupied by lizards in North America and Australia (compare Figure 11.5 with Figures 11.4 and 11.6).

To reduce dimensionality, a principal components analysis of these data was undertaken. When all 90-odd species are considered together, the first three principal components reduce overall variance by 77.7%, 11.6%, and 6.7% (total 96%), respectively. Positions of various species on these first three principal components of morphospace are depicted in Figures 11.4, 11.5, and 11.6. Centroids for each of the continental faunas (shown as small cubes) deviate from the overall centroid of all 91 species (marked by the intersections of the origins of all three principal components on Figures 11.4–11.6). Interestingly enough, even on these very crude morphological dimensions, the Australian ant specialist *Moloch horridus* (an agamid) and its American "ecological equivalent," the horned lizard *Phrynosoma platyrhinos*

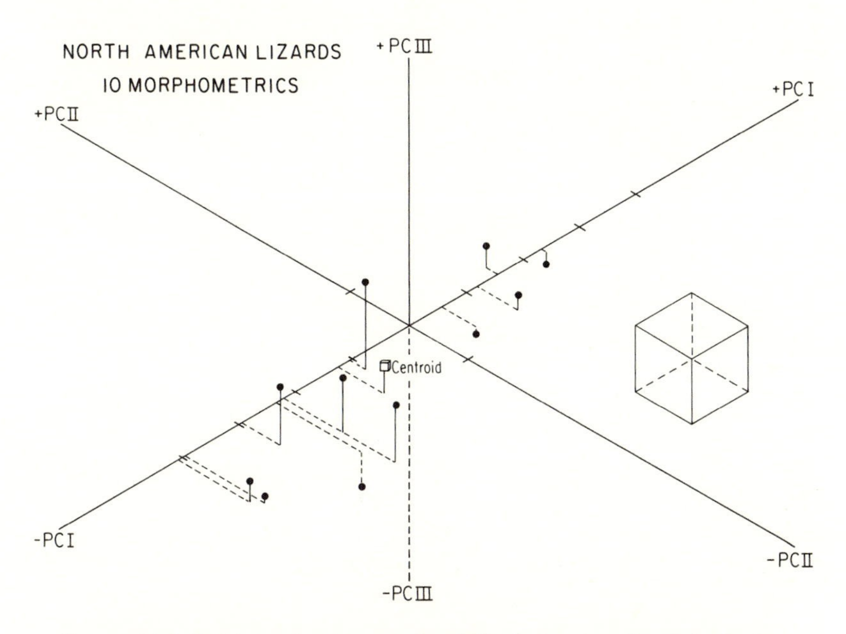


FIGURE 11.4. Positions of various species of North American lizards in the first three principal components of morphological space. Principal components are based on 91 species of lizards representing eleven families from all three continental desert-lizard systems. The centroid for the North American saurofauna is plotted as a small cube.

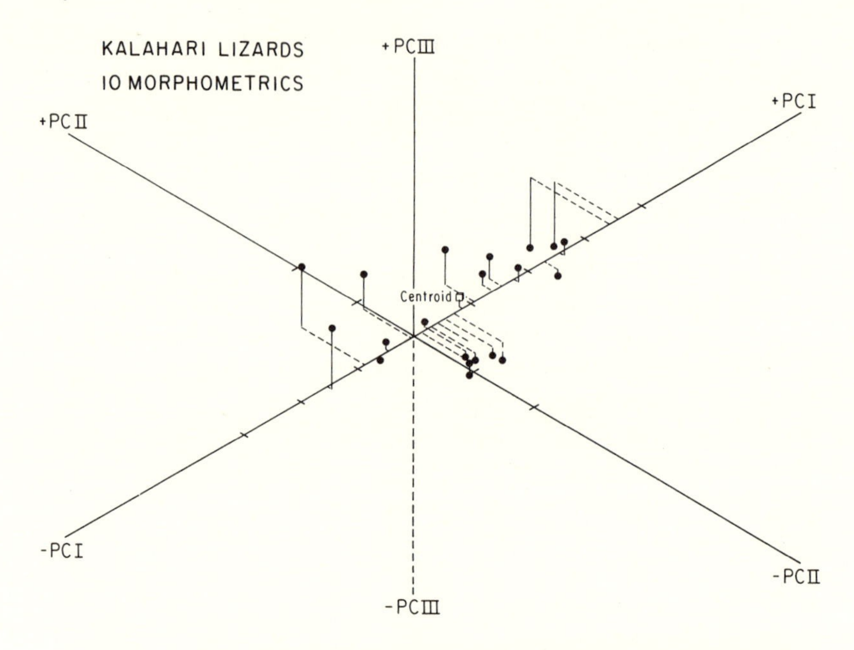


FIGURE 11.5. Positions of various species of Kalahari lizards in the first three principal components of morphological space. The centroid for Kalahari lizards is plotted as a small cube. Exactly the same coordinates are used in Figures 11.4 and 11.6 for North American and Australian lizards, respectively. Note that the morphospace occupied by Kalahari lizards is appreciably more compact than it is in the other two continental desert-lizard systems.

(an iguanid), are actually closer to one another than either is to another member of its own saurofauna.

Finally, the best available estimates of overall ecological similarity, the products of microhabitat times dietary overlap (from merged deck analyses), were compared directly to these anatomical distances. As expected, product moment correlation coefficients between ecological similarity and morphological separation are negative. Most such correlations are weak but still statistically significant (bottom line, Table 11.2). The strongest of these inverse correlations, for Kalahari lizards, is shown in Figure 11.7. Although anatomically similar species proved to be widely variable in the extent of their ecological overlap, morphologically disparate species are inevitably ecologically distinct.

The present data set can also be exploited to test the morphological variation-niche width hypothesis in at least two different ways. Estimates of microhabitat niche breadth (Appendix C) and those for di-

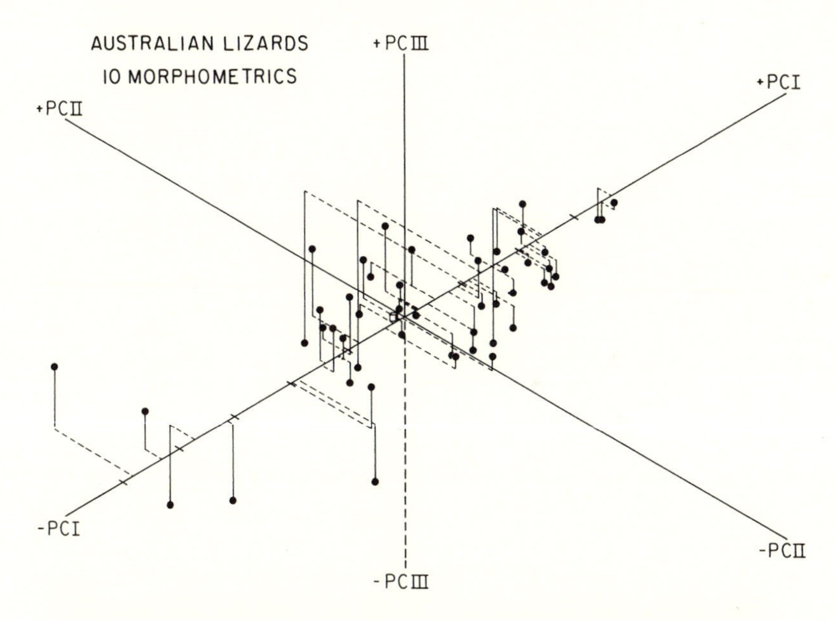


FIGURE 11.6. Positions of most species of Australian lizards in the first three principal components of morphological space (several very large species are not plotted). The centroid for Australian species (small cube) is near the overall centroid for all 91 species (origin). Coordinates for this plot are exactly identical to those used in Figures 11.4 and 11.5, with which this figure should be compared. Interestingly enough, the ant specialists *Moloch* (Australian) and *Phrynosoma* (North American) are actually closer to one another in morphological space than either species is to another member of its own saurofauna, demonstrating their ecological equivalence.

etary niche breadth (Appendix E) can be compared directly to appropriate measures of anatomical variability—namely, the coefficients of variation for hind-leg length and head length (Table 11.3). While somewhat equivocal, these tests do tend to support the hypothesis that generalists are morphologically more variable than specialists.

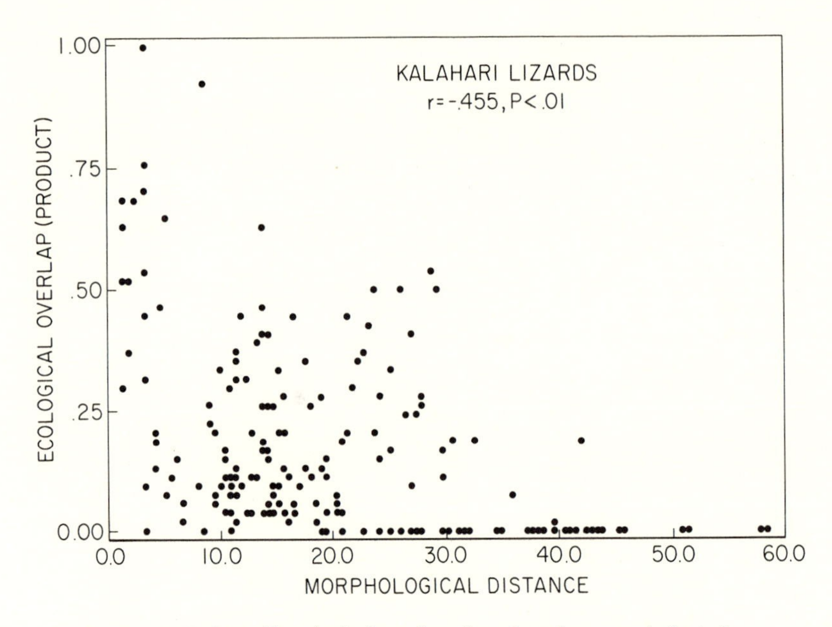


FIGURE 11.7. Overall ecological overlap plotted against morphological separation among Kalahari lizards.

TABLE 11.3. Pearson product-moment correlation coefficients between estimates of niche breadth (microhabitat, diet) and the coefficient of variation of appropriate anatomical variates (hind-leg length, head length) for various lizard assemblages.

| ASSEMBLAGE | SPATIAL | DIETARY |
|---|---|---|
| North America ($N=11$) | (.524) | −.314ns |
| Kalahari ($N=21,20$) | .553 † | .086ns |
| Australia ($N=56,57$) | .094ns | .308 * |
| All three continents ($N=88$) | .243 * | (.208) |

ns = not significant ( ) $\Sigma P \approx .05$ * = $P<.05$ † = $P<.01$

# 12 Conclusions and Prospects for Further Work

*Lack of Convergence*

Being a firm believer in the efficacy of natural selection, when I began the intercontinental aspect of these studies I fully expected the Australian deserts to be a reasonably close replica of the North American desert-lizard system, at least in terms of number of species, and perhaps even at the level of some recognizable "ecological equivalents" (Grinnell 1924). I never anticipated that I would find an exact facsimile, but imagine my total surprise as these initial expectations were dashed upon collecting a dozen different species of lizards on the first day at a good site in the Great Victoria desert! Bewildered now with more than 40 sympatric species, I never cease to marvel at the tremendous diversity of Australian desert lizards. I remain an ardent selectionist, but my views have been tempered and loosened: things are just not as tight as naive preconception might have envisioned them. The impact of history has been profound, but it is overly facile, even glib, to assert that the Australian deserts are more diverse than other deserts simply by virtue of antiquity. The Namib desert in southwestern Africa is ancient (Ward et al. 1983) and supports an extremely rich beetle fauna but yet has only a moderate diversity of lizards. Clearly, ecological factors promoting coexistence are of vital importance (Pianka 1969b, 1981). I was exceedingly fortunate to be able to find and study a desert-lizard system that is intermediate in diversity between North America and Australia—namely, the Kalahari.

Intercontinental similarities and differences are summarized in Table 12.1. Kalahari and Australian climates are warm with summer rains and are roughly comparable, although precipitation is more variable in Australia (Chapter 2, Appendix B). North American deserts en-

TABLE 12.1. Major differences and similarities among continental desert-lizard systems.

| | NORTH AMERICA |
|---|---|
| Thermal climate | Colder, shorter growing season (especially in north) |
| Precipitation | GREAT BASIN: distributed fairly evenly over year<br>MOJAVE: winter precipitation<br>SONORAN: bimodal annual rainfall, with peaks in both late summer and midwinter |
| Higher taxa (number of lizard families) | 5 |
| Number of lizard genera | 12 |
| Species diversity | |
| Lizards | Low (4–11 species) |
| Snakes | High (4.5 species) |
| Birds (all) | Low (7.8 species) |
| Insectivorous ground-feeding birds only | Low (about 2–3 species) |
| Small mammals | High (5.3 species) |
| Insect food resource diversity | High (8.7) |
| Microhabitat resource diversity | Low (2.8) |
| Lizard niche breadths | |
| Dietary | 1.1–7.3 (4.4) |
| Microhabitat | 1.0–3.9 (2.2) |
| Proportional utilization coefficients | Fit decaying exponential |
| Estimated electivities | Fit decaying exponential |
| Niche overlap | |
| Diet | Intermediate |
| Microhabitat | All-or-none |
| Overall | Highest |
| Guild structure | Little evident |
| Morphospace | Intermediate |

| KALAHARI | AUSTRALIA |
| --- | --- |
| Warmer | Warmer |
| Summer rains | Summer rain, variable annual precipitation |
| | |
| 5 | 5 |
| 13 | About 25 |
| Medium (11–18 species) | High (15–42 species) |
| Low (2.2 species) | Medium (3.6 species) |
| Medium (22.8 species) | High (28.3 species) |
| High (7.3 species) | Medium (4.8 species) |
| — | Low (1.5 species) |
| Low (4.4), dominated by termites | Intermediate (6.6) |
| High (6.8), variable | High (7.0) |
| 1.1–8.2 (3.9) | 1.0–9.4 (3.8) |
| 1.0–6.0 (3.4) | 1.0–6.4 (3.0) |
| Fit decaying exponential | Fit decaying exponential |
| Fit decaying exponential | Fit decaying exponential |
| Highest | Lowest |
| Skewed toward low values | Skewed toward low values |
| Intermediate | Lowest |
| Present | Conspicuous |
| Smallest | Largest |

compass a wider range of climatic conditions, being basically colder with a shorter growing season but a less variable precipitation regime (various subregions within North America differ substantially from one another). As indicated in Chapter 1, all three desert-lizard systems contain exactly five families of lizards, although the identities of these differ between continents. At the generic level of diversity, North America and the Kalahari are similar, but about twice as many lizard genera are recognized in Australia. Species diversities of various vertebrate taxa do not vary concordantly among continents, except for lizards and birds (but even here, subtle differences between continental desert systems can be detected; see Figure 1.5). Intercontinental differences in lizard diversity do not stem from diurnal ground-dwelling sit-and-wait foragers (absolute numbers of species of these are similar among continents), but rather arise primarily from differences in species richness of lizards that are arboreal, nocturnal, subterranean, and particularly those that are diurnal, terrestrial, and widely foraging (see Table 1.2).

Spectra of available resources, both microhabitats and foods, are broadly similar between continental systems, yet quantitative differences are evident. Surprisingly, the diversity of insect food resources is highest in North America and lowest in the Kalahari, where termites dominate lizard diets. The diversity of microhabitats used by lizards is low in North America and high but variable between sites in the Kalahari. Microhabitat and dietary niche breadths do not differ appreciably between continents, with the exception that microhabitat niche breadths are slightly broader in the Kalahari than in North America (Chapter 3). On all three continents, both proportional utilization coefficients and estimated electivities are strongly skewed toward low values and approximate decaying exponentials. Dietary overlap is highest in the Kalahari due to the heavy consumption of termites, lowest in Australia, and intermediate in North America. Microhabitat overlap is essentially all-or-none in North America but skewed toward low values in the two southern hemisphere deserts. Overall niche overlap, estimated as the product of dietary times microhabitat overlap, is lowest in Australia and highest in North America. Relatively little guild structure is evident in North America, attributable in part to the low diversity of these systems; but guilds are clearly present in the Kalahari and are conspicuous in Australia. The morphological hypervolume occupied by lizards is largest in Australia as might be expected, but most compact in the Kalahari where most species of lizards are small.

*Prospects for Further Work*

The more I learn about independently evolved desert-lizard systems the less confident I become about what I do in fact "know."[1] As others have pointed out before, increasing scientific knowledge only serves to heighten our awareness of the frontiers of ignorance. Thus efforts to understand reality inevitably become almost counterproductive. On a smaller scale, my original intent in writing this monograph was to try to tie up all loose ends and to present a unified coherent picture; as analysis and writing progressed, new questions and avenues for further work kept arising so that the "final" product is far from finished. Perhaps all symphonies are doomed to remain unfinished! The full story as to why these continental desert-lizard systems differ so much will clearly never be known. Even so, directions for further research would seem appropriate to consider at this point.

The present data base can no doubt be exploited in many more ways to enhance understanding of these desert-lizard systems. For example, null hypotheses need to be developed for the alien-versus-resident comparisons made in Chapter 9. An analogous analysis needs to be undertaken that does not assume competitive communities in equilibria and the Lotka-Volterra competition equations. Much more could be done with niche breadths. Work on modeling and understanding events involved in the fire succession cycle (pp. 133–134), as well as new efforts to collect data relevant to it, could also prove to be most fruitful. It might even be feasible to exploit controlled burns to test some aspects of fire succession models.

Still another obvious suggestion is to attempt direct experimental manipulations of the abundances of component species (Schoener 1983). Manipulations of densities are sometimes possible, if exceedingly tedious and difficult, as long as relatively small numbers of species are involved. For example, such reciprocal removal experiments proved feasible with two saxicolous species of lizards in western Texas because rock outcrops formed isolated habitat islands (Dunham 1980). However, the very vastness and continuity that make flatland desert systems so amenable to observation and sampling and thus such tractable study sites render removal experiments next to impossible. Aus-

[1] A curious attitude I have sometimes encountered is that these lizards have now been studied so thoroughly that little of interest will emerge from further study. Nothing could be farther from the truth. Perhaps I have packaged them too prettily and "sold" them too effectively! A great deal remains to be learned about even the common species, let alone the rarer ones, many of which are virtually unknown. As noted above, one of the most mystifying problems is how such exceedingly uncommon species can even manage to continue to exist.

tralian desert lizards simply do not lend themselves to manipulative studies. One must often work all day to collect just a few dozen animals; moreover, considerable expertise is required just to find some species. For some of the rarer species, a hundred or more man-days must be spent in the deserts before encountering even a single individual. Even if removals were feasible, the sheer number of pairwise combinations would certainly preclude complete analysis of the most interesting and diverse systems. Furthermore, as pointed out by Bender et al. (1984), results from such experiments are ambiguous because they cannot distinguish direct pairwise effects from more indirect ones of the sort envisaged by Lawlor (1979) and others. Experiments obviously will have to be thought out and set up with great care. Clearly, the challenge of these diverse communities remains awesome.

Observations, on other, independently evolved extant desert lizard systems would also be of obvious comparative interest. However, it is dubious that any can even begin to compete with the Australian deserts for sheer diversity. The other major North American desert, the Chihuahuan, is high in elevation and rather low in lizard diversity (Maury 1981; Maury and Barbault 1981). In South America, the Argentine Monte and the Peruvian Atacama (including the Sechura and others) have been examined to a limited extent: both are very low in lizard diversity (Orians and Solbrig 1977; Huey 1979; Pearson and Ralph 1978; Sage 1972, 1974). As indicated above, the ancient and very arid Namib desert of southwestern Africa supports a moderate diversity of interesting endemic lizard species (particularly geckos and lacertids), but most have very restricted geographic distributions, and diversity at any one point is invariably low (Haacke 1976; Werner 1977; Robinson and Cunningham 1978). A Mediterranean lizard community in Spain consists of seven species of lizards plus one amphisbaenid, which differ in microhabitats and diets (Mellado et al. 1975). In northern Africa, the Sahara and adjacent desert regions house some fascinating but largely unstudied lizard species, as do the Arabian and mid-eastern deserts. The high-latitude cold deserts of central Asia, including the Gobi, also unfortunately little-known to the western world (Lowe 1968), are presumably low in diversity as well.

Comparative ecology has proven to be largely a descriptive rather than a predictive science. I would have liked to end this discussion by making some testable predictions about unknown desert saurofaunas (not too many more desert-lizard systems exist), based upon what has been gleaned from the studies detailed here. Unfortunately, the sorts of predictions possible are somewhat trivial and limited. For example, the Gobi desert is at a high latitude and quite cold with a short growing season. It should thus be most comparable to the North American Great Basin desert. Hence one expects Gobi desert lizards to be diurnal heliotherms that thermoregulate actively with a low slope and high intercept on plots of body temperature versus air temperature. Thus their clutch sizes would probably not be small (see p. 66). One would not expect to find nocturnal lizards, except perhaps at very southern localities (only one species of gecko, *Teratoscincus przewalski*, seems to occur in the region, and it may very well be diurnal). Further, one might expect to find a widely foraging species comparable to *Cnemidophorus tigris* (this would probably be a lacertid in the genus *Eremias*) plus several sit-and-wait foragers (these might possibly be other *Eremias* but could perhaps be agamids such as *Phrynocephalus*). One of these might be expected to be an ant-specialized analogue of *Phrynosoma* and *Moloch*, with a spiny tanklike body form (the best candidate for this role would probably be a *Phrynocephalus*). A small collection of preserved Gobi desert lizards taken at one site, with supporting data on body temperatures and air temperatures, would obviously enable much more definitive "predictions."

The importance and functional significance of nonlizard components in these desert systems obviously needs to be assessed and studied in considerably greater detail. Resource dynamics, as well as populations of predators and potential competing consumers other than lizards, may prove to be particularly crucial to evaluate. Moreover, the approach adopted here would benefit immeasurably from being extended to include all species of desert lizards—or perhaps even better, all known lizard species. Pseudo-community subsets drawn from such worldwide species pools might well allow penetrating insights to be drawn into existing assemblages (for an avian example, see Schoener 1984). When such data are finally coupled with complete systematic information on probable phylogenetic relationships and biogeographic events, our understanding and appreciation of the complex impact of historical constraints in time and space on the assembly of lizard communities and the evolution of lizard ecology will be greatly enhanced.

Somehow, it is strangely comforting to me to know that, long after we humans have had our day, 40-plus species of lizards will no doubt still roam free in the Great Victoria desert, interacting with one another as they have for millennia, blissfully oblivious of our modest efforts to understand those interactions.

# Appendices

# Appendix A
# Lizard Censuses and Relative Abundances

Appendix A.1. Lizard censuses from twelve study areas in the North American desert system.

| | Great Basin Desert | | | Mojave Desert | | | | Sonoran Desert | | | | |
|---|---|---|---|---|---|---|---|---|---|---|---|---|
| Species/Areas | I | L | G | V | S | P | M | T | W | C | A | B |
| *Cnemidophorus tigris* | 84 | 85 | 64 | 80 | 189 | 91 | 499 | 186 | 137 | 467 | 28 | 70 |
| *Uta stansburiana* | 3 | 95 | 37 | 40 | 64 | 20 | 150 | 22 | 60 | 361 | 44 | 25 |
| *Crotaphytus wislizeni* | 11 | 56 | 40 | 15 | 5 | 4 | 4 | 14 | 2 | 3 | 6 | |
| *Phrynosoma platyrhinos* | 2 | 38 | 15 | 11 | 8 | 1 | 38 | 9 | 3 | 15 | | 1 |
| *Callisaurus draconoides* | | 22 | 1 | 81 | 2 | 33 | 32 | 26 | 15 | 49 | 96 | 46 |
| *Sceloporus magister* | | | | | | | 7 | | 9 | 16 | 19 | 26 |
| *Sceloporus clarki* | | | | | | | | | | | 3 | |
| *Urosaurus graciosus* | | | | | | | | 5 | 4 | 25 | | 1 |
| *Urosaurus ornatus* | | | | | | | | | | | 66 | 36 |
| *Dipsosaurus dorsalis* | | | | | | 3 | | 15 | 6 | 21 | | 10 |
| *Uma scoparia* | | | | | | | | 32 | | | | |
| *Xantusia vigilis* | | | | | | | 21 | | | | | |
| *Coleonyx variegatus* | | | | 3 | 2 | 1 | 3 | 5 | 9 | 18 | | 3 |
| *Heloderma suspectum* | | | | | | | | | | 1 | | |
| Total number of lizards | 100 | 296 | 157 | 230 | 270 | 153 | 754 | 314 | 245 | 976 | 262 | 218 |
| Total number of species | 4 | 5 | 5 | 6 | 6 | 7 | 8 | 9 | 9 | 10 | 7 | 9 |
| Species diversity | 1.4 | 4.1 | 3.4 | 3.6 | 1.8 | 2.4 | 2.1 | 2.6 | 2.6 | 2.7 | 4.1 | 4.9 |

Appendix A.2. Lizard censuses from ten study sites in the Kalahari semidesert.

| Species/Areas | A | B | D | G | K | L | M | R | T | X |
|---|---|---|---|---|---|---|---|---|---|---|
| *Agama hispida* | 37 | 44 | 31 | 37 | 27 | 29 | 1 | 15 | 11 | 67 |
| *Chameleo dilepis* | | | | | | | | | 2 | |
| *Eremias lineo-ocellata* | 178 | 95 | 123 | 159 | 48 | 69 | 43 | 19 | 140 | 157 |
| *Eremias lugubris* | 5 | | 13 | 13 | 5 | 44 | 2 | 4 | 125 | 21 |
| *Eremias namaquensis* | | 19 | | 10 | 80 | 61 | | 8 | 34 | 4 |
| *Ichnotropis squamulosa* | | | | 5 | | | | | 110 | |
| *Meroles suborbitalis* | 160 | 25 | 31 | 72 | 3 | 187 | 4 | 135 | | 85 |
| *Nucras intertexta* | | | | | | | | | 3 | |
| *Nucras tessellata* | 1 | 26 | | | | 12 | | | | |
| *Mabuya occidentalis* | 32 | 29 | 15 | 19 | 40 | 26 | 2 | 6 | 29 | 5 |
| *Mabuya spilogaster* | | | 1 | | 96 | | | | 222 | |
| *Mabuya striata* | 16 | 128 | | 11 | 134 | 66 | 14 | 7 | | 83 |
| *Mabuya variegata* | 64 | 8 | 3 | 4 | | | 5 | | 3 | 21 |
| *Typhlosaurus gariepensis* | 22 | 70 | | | 2 | 6 | 1 | | | 8 |
| *Typhlosaurus lineatus* | 30 | 32 | 10 | | 74 | 132 | 26 | 4 | 104 | 30 |
| *Chondrodactylus angulifer* | 66 | 46 | 48 | 49 | 33 | 44 | 26 | 7 | 71 | 10 |
| *Colopus wahlbergi* | 6 | 4 | 3 | 7 | 8 | 4 | 4 | | 74 | 5 |
| *Lygodactylus capensis* | | 18 | | | | | | | | |
| *Pachydactylus bibroni* | | 32 | | | 38 | 1 | | | 1 | 5 |
| *Pachydactylus capensis* | 7 | 3 | 3 | 6 | | 1 | 1 | 9 | 34 | 1 |
| *Pachydactylus rugosus* | 1 | 3 | 4 | | 2 | 2 | 4 | | | 8 |
| *Ptenopus garrulus* | 14 | 3 | 83 | 17 | 48 | 6 | 18 | 6 | 182 | 20 |
| Total number of lizards | 639 | 585 | 368 | 409 | 638 | 690 | 151 | 220 | 1145 | 530 |
| Total number of species | 15 | 17 | 13 | 13 | 15 | 16 | 14 | 11 | 16 | 16 |
| Species diversity | 5.8 | 8.7 | 5.1 | 4.7 | 8.4 | 6.7 | 6.0 | 2.5 | 8.5 | 6.1 |

Appendix A.3. Lizard censuses from nine study areas in Western Australia.

| Species/Areas | A | D | E | G | $L_1$ | $L_2$ | M | N | R | Y |
|---|---|---|---|---|---|---|---|---|---|---|
| *Caimanops amphiboluroides* | 13 | | | | | | | | | |
| *Ctenophorus clayi* | | | 1 | | | | | | 16 | |
| *Ctenophorus fordi* | | | 105 | | | | | | 1 | |
| *Ctenophorus inermis* | 47 | 16 | 11 | 6 | 9 | 5 | 1 | 2 | 15 | 2 |
| *Ctenophorus isolepis* | | 93 | 82 | 113 | 55 | 530 | 76 | 18 | 209 | 9 |
| *Ctenophorus reticulatus* | | | | | | | | | | 14 |
| *Ctenophorus scutulatus* | 74 | | | | | | 1 | | | 1 |
| *Diporiphora winneckei* | | 18 | 17 | | | | | | | |
| *Gemmatophora longirostris* | | 29 | 11 | | | | | | 26 | |
| *Moloch horridus* | 3 | 12 | 44 | 1 | 9 | 9 | 2 | | 70 | |
| *Pogona minor* | 14 | 1 | 9 | 3 | 8 | 12 | 4 | | 10 | |
| *Varanus brevicauda* | | 1 | | | | | 1 | | | |
| *Varanus caudolineatus* | 12 | | | | | | | | | |
| *Varanus eremius* | 13 | 10 | 14 | 3 | 3 | 13 | 2 | 1 | 7 | |
| *Varanus gilleni* | | | | | | | | | 1 | |
| *Varanus giganteus* | | | | | | | | | 1 | |
| *Varanus gouldi* | 2 | 1 | 8 | 1 | tr* | 4 | 1 | 1 | 5 | 1 |
| *Varanus tristis* | (1)tr | 3 | 3 | 2 | 6 | 18 | 1 | | 38 | |
| *Ctenotus ariadnae* | | | | | 5 | 11 | | | 1 | |
| *Ctenotus atlas* | 23 | | | | | | | | | |
| *Ctenotus brooksi* | | | 28 | | | | | | 4 | |
| *Ctenotus calurus* | | 10 | 4 | 3 | 11 | 147 | 1 | 5 | 14 | |
| *Ctenotus colletti* | | | 3 | | | | | | 8 | |
| *Ctenotus dux* | | 9 | 36 | | | | | | 112 | |
| *Ctenotus grandis* | 7 | | 1 | 2 | 45 | 39 | 1 | 3 | 5 | |
| *Ctenotus helenae* | | 2 | 2 | 6 | 53 | 31 | 5 | 1 | 26 | |
| *Ctenotus leae* | | | 18 | | | | | | | |
| *Ctenotus leonhardii* | 86 | | | | | | 6 | | | 3 |
| *Ctenotus pantherinus* | 5 | 2 | 10 | 17 | 8 | 21 | 16 | 2 | 25 | |
| *Ctenotus piankai* | | | 1 | 2 | 2 | 3 | | 5 | 8 | |
| *Ctenotus quattuordecimlineatus* | | | 7 | 31 | 46 | 183 | | 1 | 90 | |
| *Ctenotus schomburgkii* | 53 | | 8 | 6 | 7 | 40 | 11 | | | 2 |
| *Cryptoblepharus plagiocephalus* | 24 | | | 2 | 27 | 53 | | | | |
| *Egernia depressa* | 12 | | | | | | | | | |
| *Egernia kintorei* | | | | | 1 | 0 | | | | |
| *Egernia inornata* | 10 | | 4 | 7 | 2 | 2 | | 6 | 9 | 9 |
| *Egernia striata* | | 4 | 6 | 10 | 37 | 68 | 24 | 4 | 13 | |
| *Lerista bipes* | | | 4 | 2 | 6 | 53 | | | 5 | |
| *Lerista desertorum* | | | 1 | | | | | | 1 | |

Appendix A.3. *(continued)*

| Species/Areas | A | D | E | G | $L_1$ | $L_2$ | M | N | R | Y |
|---|---|---|---|---|---|---|---|---|---|---|
| *Lerista muelleri* | 3 | | | | 0 | 6 | | | | |
| *Menetia greyi* | 3 | | 1 | | 3 | 4 | | 3 | 1 | |
| *Morethia butleri* | 6 | | 2 | 1 | 0 | 1 | | | 1 | |
| *Omolepida branchialis* | | | | 2 | | | | | 3 | |
| *Eremiascincus richardsoni* | | | | | | | | | 2 | 5 |
| *Tiliqua multifasciata* | | | 1 | | | | | | | |
| *Delma fraseri* | 8 | | | | 0 | 4 | | | 3 | |
| *Lialis burtonis* | 1 | | 1 | | 2 | 2 | 1 | | 2 | |
| *Pygopus nigriceps* | | 1 | | | 8 | 3 | | | 3 | |
| *Diplodactylus ciliaris* | | | 21 | | | | 34 | | 43 | |
| *Diplodactylus conspicillatus* | | 6 | 4 | 9 | 6 | 27 | 24 | | 10 | 1 |
| *Diplodactylus damaeus* | | 31 | 8 | | | | | | 20 | |
| *Diplodactylus elderi* | 12 | 3 | 1 | | 0 | 4 | | | 13 | |
| *Diplodactylus pulcher* | 16 | | | | | | | | | |
| *Diplodactylus stenodactylus* | | | | 1 | 0 | 1(2)† | | | | |
| *Diplodactylus strophurus* | 30 | 11 | 2 | | | | | | 12 | 3 |
| *Gehyra variegata* | 32 | 8 | 24 | 48 | 114 | 202 | 8 | | 378 | 2 |
| *Heteronotia binoei* | 6 | | 1 | 1 | 1 | 1 | | 3 | 1 | 3 |
| *Nephrurus laevissimus* | | 82 | 85 | | | | | | 193 | |
| *Nephrurus levis* | | | | 2 | 6 | 4 | | 8 | | |
| *Nephrurus vertebralis* | | | | | | | 3 | | | 10 |
| *Rhynchoedura ornata* | 132 | 7 | 10 | 28 | 50 | 64 | 12 | 10 | 31 | 1 |
| Total number of lizards | 648 | 360 | 599 | 309 | 530 | 1565 | 235 | 73 | 1436 | 66 |
| Total number of species | 28 | 23 | 39 | 26 | 27 | 32 | 22 | 16 | 42 | 15 |
| Species diversity | 10.5 | 7.0 | 11.3 | 5.5 | 9.9 | 6.2 | 6.3 | 8.5 | 8.0 | 8.3 |

*Tracks

†One collected, one found inside stomach of another lizard.

# Appendix B
## Annual Precipitation Statistics

Appendix B. Annual precipitation statistics from weather stations closest to various study areas.

| Continent/Study Area | | Mean | Standard Deviation | Coefficient of Variation | Autocorrelation Coefficient |
|---|---|---|---|---|---|
| NORTH AMERICA | | | | | |
| N=I | Idaho | 18.4 | 5.6 | .304 | +0.32 |
| L | Lovelock | 14.0 | 5.5 | .393 | +0.13 |
| G | Gabbs | 9.6 | 4.4 | .458 | −0.17 |
| V | Grapevine Canyon | 11.6 | 6.6 | .569 | −0.59 |
| S | Searchlight | 18.7 | 10.5 | .562 | −0.14 |
| P | Pahrump | 9.3 | 5.9 | .634 | −0.52 |
| M=Y | Mojave | 12.7 | 9.2 | .724 | +0.07 |
| T=O | 29 Palms | 9.4 | 6.1 | .649 | −0.19 |
| R=W | Salome | 19.2 | 8.0 | .417 | +0.08 |
| C=B | Casa Grande | 20.9 | 7.6 | .364 | −0.05 |
| KALAHARI | | | | | |
| R | Rhigozum | 22.7 | 11.3 | .498 | −0.38 |
| D | Farm D | 21.7 | 10.3 | .475 | +0.01 |
| G | Geselskop | 22.7 | 11.3 | .498 | −0.38 |
| A | Aarpan | 14.5 | 8.6 | .593 | −0.08 |
| L | Ludrille | 16.7 | 7.8 | .467 | +0.08 |
| X | Farm X | 19.0 | 9.7 | .511 | −0.27 |
| T | Tsabong | 28.6 | 9.3 | .325 | −0.03 |
| B | Bloukrans | 15.2 | 7.2 | .474 | +0.14 |
| AUSTRALIA | | | | | |
| M | Millrose | 21.5 | 13.4 | .623 | −0.14 |
| D | Lorna Glen | 23.5 | 14.0 | .596 | −0.03 |
| A | Atley HS | 21.3 | 10.0 | .470 | −0.14 |
| L | Laverton | 21.9 | 9.3 | .425 | +0.01 |
| E | E-area | 20.2 | 13.1 | .649 | −0.23 |
| R | Red Sands | 31.2 | 17.7 | .567 | +0.36 |

Note: Weather records were not available for several remote sites.

# Appendix C
# Microhabitat Resource Matrices and Niche Breadths

Appendix C.1. Percentage utilization of various microhabitats among various North American species of desert lizards, with total sample sizes.

| | TERRESTRIAL | | | | | TERRESTRIAL | | | | |
|---|---|---|---|---|---|---|---|---|---|---|
| Species | *Open Sun* | *Grass Sun* | *Bush Sun* | *Tree Sun* | *Other Sun* | *Open Shade* | *Grass Shade* | *Bush Shade* | *Tree Shade* | *Other Shade* |
| *Cnemidophorus tigris* | 47.5 | 2.5 | 34.6 | 2.5 | 2.5 | 0.0 | 0.0 | 7.8 | 0.0 | 2.5 |
| *Uta stansburiana* | 27.1 | 4.6 | 41.3 | 4.6 | 4.6 | 0.8 | 0.0 | 6.0 | 0.0 | 3.9 |
| *Phrynosoma platyrhinos* | 92.6 | 0.0 | 2.5 | 0.0 | 0.0 | 2.5 | 0.0 | 2.5 | 0.0 | 0.0 |
| *Crotaphytus wislizeni* | 59.1 | 0.7 | 22.6 | 0.7 | 0.7 | 2.2 | 0.0 | 9.5 | 0.0 | 0.0 |
| *Callisaurus draconoides* | 80.7 | 2.0 | 4.3 | 3.9 | 2.0 | 0.7 | 0.4 | 0.9 | 0.4 | 0.4 |
| *Sceloporus magister* | 0.0 | 0.0 | 1.2 | 1.2 | 3.5 | 0.0 | 0.0 | 3.5 | 1.2 | 1.2 |
| *Urosaurus graciosus* | 0.0 | 0.0 | 2.2 | 0.0 | 0.0 | 0.0 | 0.0 | 0.0 | 0.0 | 0.0 |
| *Dipsosaurus dorsalis* | 46.3 | 1.5 | 40.3 | 1.5 | 1.5 | 0.0 | 0.0 | 9.0 | 0.0 | 0.0 |
| *Uma scoparia* | 80.5 | 0.8 | 7.3 | 0.8 | 0.8 | 2.4 | 0.0 | 7.3 | 0.0 | 0.0 |
| *Coleonyx variegatus* | 0.0 | 0.0 | 0.0 | 0.0 | 0.0 | 95.3 | 0.0 | 4.7 | 0.0 | 0.0 |
| *Xantusia vigilis* | 0.0 | 0.0 | 0.0 | 0.0 | 0.0 | 0.0 | 0.0 | 0.0 | 100.0 | 0.0 |
| Total number of lizards | 1655 | 94 | 1029 | 105 | 97 | 58.5 | 2 | 221 | 30 | 77.5 |

| ARBOREAL | | | | | |
|---|---|---|---|---|---|
| *Low Sun* | *Low Shade* | *High Sun* | *High Shade* | *Sample Size, N* | *Microhabitat Niche Breadth* |
| 0.1 | 0.0 | 0.0 | 0.0 | 1801 | 2.82 |
| 5.3 | 0.1 | 1.8 | 0.0 | 768 | 3.87 |
| 0.0 | 0.0 | 0.0 | 0.0 | 121 | 1.17 |
| 2.2 | 0.0 | 2.2 | 0.0 | 137 | 2.43 |
| 3.2 | 0.0 | 1.1 | 0.0 | 538 | 1.52 |
| 10.0 | 5.9 | 38.2 | 34.1 | 85 | 3.58 |
| 7.7 | 1.1 | 39.6 | 49.5 | 46 | 2.45 |
| 0.0 | 0.0 | 0.0 | 0.0 | 67 | 2.60 |
| 0.0 | 0.0 | 0.0 | 0.0 | 41 | 1.52 |
| 0.0 | 0.0 | 0.0 | 0.0 | 43 | 1.10 |
| 0.0 | 0.0 | 0.0 | 0.0 | 27 | 1.00 |
| 73.5 | 6 | 73.5 | 51.5 | 3674 | 2.19 (Mean) |

Appendix C.2. Percentage utilization of various microhabitats among various species of Kalahari desert lizards, with total sample sizes.

| Species | Subt. | Terrestrial | | | | | Terrestrial | | | | |
|---|---|---|---|---|---|---|---|---|---|---|---|
| | | *Open Sun* | *Grass Sun* | *Bush Sun* | *Tree Sun* | *Other Sun* | *Open Shade* | *Grass Shade* | *Bush Shade* | *Tree Shade* | *Other Shade* |
| *Agama hispida* | 0.0 | 46.7 | 1.1 | 15.4 | 2.1 | 0.0 | 4.7 | 0.4 | 11.8 | 2.5 | 0.0 |
| *Chameleo dilepis* | 0.0 | 50.0 | 0.0 | 0.0 | 0.0 | 0.0 | 0.0 | 0.0 | 0.0 | 0.0 | 0.0 |
| *Eremias lineo-ocellata* | 0.0 | 30.7 | 10.8 | 21.9 | 3.7 | 0.0 | 1.5 | 6.8 | 21.2 | 2.7 | 0.3 |
| *Eremias lugubris* | 0.0 | 35.1 | 6.6 | 29.0 | 5.2 | 0.2 | 0.7 | 2.4 | 17.2 | 3.3 | 0.2 |
| *Eremias namaquensis* | 0.0 | 53.0 | 1.2 | 24.5 | 2.4 | 0.0 | 5.2 | 0.6 | 11.8 | 1.2 | 0.0 |
| *Ichnotropis squamulosa* | 0.0 | 18.3 | 17.2 | 22.6 | 0.5 | 0.0 | 1.1 | 16.1 | 18.3 | 5.9 | 0.0 |
| *Meroles suborbitalis* | 0.0 | 56.8 | 3.1 | 23.2 | 0.0 | 0.0 | 2.2 | 1.8 | 12.5 | 0.0 | 0.2 |
| *Nucras intertexta* | 0.0 | 0.0 | 16.7 | 33.3 | 0.0 | 0.0 | 0.0 | 50.0 | 0.0 | 0.0 | 0.0 |
| *Nucras tessellata* | 0.0 | 30.7 | 4.0 | 32.7 | 0.0 | 0.0 | 2.7 | 6.7 | 22.0 | 1.3 | 0.0 |
| *Mabuya occidentalis* | 0.0 | 4.1 | 2.7 | 28.4 | 0.7 | 0.5 | 1.2 | 8.5 | 38.6 | 0.7 | 1.0 |
| *Mabuya spilogaster* | 0.0 | 6.2 | 0.0 | 2.1 | 11.6 | 0.0 | 0.8 | 0.0 | 2.1 | 11.1 | 0.0 |
| *Mabuya striata* | 0.0 | 1.8 | 0.5 | 2.3 | 10.1 | 1.2 | 1.3 | 0.5 | 2.3 | 10.0 | 1.0 |
| *Mabuya variegata* | 0.0 | 2.7 | 24.7 | 5.3 | 0.0 | 0.0 | 2.7 | 59.3 | 5.3 | 0.0 | 0.0 |
| *Typhlosaurus gariepensis* | 100.0 | 0.0 | 0.0 | 0.0 | 0.0 | 0.0 | 0.0 | 0.0 | 0.0 | 0.0 | 0.0 |
| *Typhlosaurus lineatus* | 100.0 | 0.0 | 0.0 | 0.0 | 0.0 | 0.0 | 0.0 | 0.0 | 0.0 | 0.0 | 0.0 |
| *Chondrodactylus angulifer* | 0.0 | 0.0 | 0.0 | 0.0 | 0.0 | 0.0 | 35.7 | 16.5 | 41.0 | 5.5 | 0.5 |
| *Colopus wahlbergi* | 0.0 | 0.0 | 0.0 | 0.0 | 0.0 | 0.0 | 43.4 | 17.2 | 27.9 | 11.5 | 0.0 |
| *Lygodactylus capensis* | 0.0 | 0.0 | 0.0 | 0.0 | 0.0 | 0.0 | 0.0 | 0.0 | 0.0 | 0.0 | 0.0 |
| *Pachydactylus bibroni* | 0.0 | 0.0 | 0.0 | 0.0 | 0.0 | 0.0 | 3.3 | 0.0 | 3.3 | 2.2 | 1.1 |
| *Pachydactylus capensis* | 0.0 | 0.0 | 0.0 | 0.0 | 0.0 | 0.0 | 9.2 | 6.2 | 27.7 | 12.3 | 3.1 |
| *Pachydactylus rugosus* | 0.0 | 0.0 | 0.0 | 0.0 | 0.0 | 0.0 | 4.5 | 4.5 | 9.1 | 4.5 | 0.0 |
| *Ptenopus garrulus* | 0.0 | 0.0 | 0.0 | 0.0 | 0.0 | 0.0 | 64.8 | 7.3 | 23.9 | 3.1 | 0.9 |
| Total number of lizards | 579 | 890 | 155 | 547 | 126 | 6 | 546 | 274 | 765 | 179 | 18 |

Appendix C.3. Percentage utilization of various microhabitats among various species of Australian desert lizards, with total sample sizes.

| Species | Subt. | Terrestrial | | | | | Terrestrial | | | | |
|---|---|---|---|---|---|---|---|---|---|---|---|
| | | *Open Sun* | *Grass Sun* | *Bush Sun* | *Tree Sun* | *Other Sun* | *Open Shade* | *Grass Shade* | *Bush Shade* | *Tree Shade* | *Other Shade* |
| *Caimanops amphiboluroides* | 0.0 | 6.3 | 18.8 | 0.0 | 0.0 | 0.0 | 6.3 | 0.0 | 6.3 | 0.0 | 0.0 |
| *Ctenophorus clayi* | 0.0 | 45.7 | 17.4 | 8.7 | 0.0 | 0.0 | 10.9 | 8.7 | 4.3 | 0.0 | 0.0 |
| *Ctenophorus fordi* | 0.0 | 50.9 | 3.6 | 15.8 | 0.0 | 0.9 | 3.6 | 2.7 | 21.6 | 0.0 | 0.9 |
| *Ctenophorus inermis* | 0.0 | 30.1 | 21.9 | 0.6 | 3.9 | 0.0 | 2.0 | 6.2 | 0.0 | 1.1 | 0.0 |
| *Ctenophorus isolepis* | 0.0 | 35.8 | 45.0 | 4.9 | 0.3 | 0.0 | 1.8 | 8.8 | 3.2 | 0.2 | 0.0 |
| *Ctenophorus reticulatus* | 0.0 | 28.1 | 0.0 | 6.3 | 0.0 | 0.0 | 6.3 | 3.1 | 6.3 | 0.0 | 0.0 |
| *Ctenophorus scutulatus* | 0.0 | 32.5 | 0.0 | 14.9 | 3.9 | 0.0 | 2.6 | 0.0 | 34.4 | 3.9 | 0.0 |
| *Diporiphora winneckei* | 0.0 | 5.7 | 10.0 | 38.6 | 0.0 | 1.4 | 0.0 | 1.4 | 10.0 | 0.0 | 1.4 |
| *Gemmatophora longirostris* | 0.0 | 7.8 | 4.5 | 38.3 | 1.3 | 0.6 | 2.6 | 1.9 | 7.1 | 3.9 | 0.6 |
| *Moloch horridus* | 0.0 | 23.0 | 7.9 | 5.8 | 0.0 | 1.2 | 4.8 | 21.0 | 28.2 | 3.6 | 2.8 |

| ARBOREAL | | | | | |
|---|---|---|---|---|---|
| *Low Sun* | *Low Shade* | *High Sun* | *High Shade* | *Sample Size, N* | *Microhabitat Niche Breadth* |
| 3.0 | 1.2 | 5.4 | 5.8 | 285 | 3.76 |
| 0.0 | 0.0 | 0.0 | 50.0 | 2 | 2.00 |
| 0.0 | 0.3 | 0.0 | 0.1 | 688 | 4.86 |
| 0.0 | 0.0 | 0.0 | 0.0 | 212 | 4.06 |
| 0.0 | 0.0 | 0.0 | 0.0 | 165 | 2.79 |
| 0.0 | 0.0 | 0.0 | 0.0 | 93 | 5.65 |
| 0.2 | 0.0 | 0.0 | 0.0 | 538 | 2.54 |
| 0.0 | 0.0 | 0.0 | 0.0 | 3 | 2.57 |
| 0.0 | 0.0 | 0.0 | 0.0 | 75 | 3.91 |
| 2.1 | 2.1 | 4.7 | 4.7 | 206 | 4.08 |
| 14.2 | 3.8 | 24.0 | 24.0 | 328 | 5.96 |
| 16.4 | 3.8 | 23.8 | 24.9 | 390 | 5.91 |
| 0.0 | 0.0 | 0.0 | 0.0 | 75 | 2.38 |
| 0.0 | 0.0 | 0.0 | 0.0 | 125 | 1.00 |
| 0.0 | 0.0 | 0.0 | 0.0 | 454 | 1.00 |
| 0.0 | 0.4 | 0.0 | 0.4 | 417 | 3.07 |
| 0.0 | 0.0 | 0.0 | 0.0 | 122 | 3.23 |
| 2.6 | 31.6 | 13.2 | 52.6 | 19 | 2.53 |
| 0.0 | 41.8 | 0.0 | 48.4 | 91 | 2.43 |
| 0.0 | 29.2 | 0.0 | 12.3 | 65 | 4.86 |
| 0.0 | 34.1 | 0.0 | 43.2 | 22 | 3.15 |
| 0.0 | 0.0 | 0.0 | 0.0 | 423 | 2.07 |
| 125 | 110 | 201 | 277 | 4798 | 3.36 (Mean) |

| ARBOREAL | | | | | |
|---|---|---|---|---|---|
| *Low Sun* | *Low Shade* | *High Sun* | *High Shade* | *Sample Size, N* | *Microhabitat Niche Breadth* |
| 0.0 | 0.0 | 62.5 | 0.0 | 16 | 2.29 |
| 4.3 | 0.0 | 0.0 | 0.0 | 23 | 3.71 |
| 0.0 | 0.0 | 0.0 | 0.0 | 111 | 2.99 |
| 8.4 | 1.4 | 19.1 | 5.3 | 178 | 5.24 |
| 0.1 | 0.0 | 0.0 | 0.0 | 1172 | 2.93 |
| 15.6 | 3.1 | 21.9 | 9.4 | 32 | 5.75 |
| 3.9 | 2.6 | 0.0 | 1.3 | 77 | 3.96 |
| 17.1 | 8.6 | 5.7 | 0.0 | 35 | 4.70 |
| 8.4 | 7.1 | 9.1 | 6.5 | 77 | 5.33 |
| 0.8 | 0.8 | 0.0 | 0.0 | 124 | 5.24 |

| SPECIES | Subt. | TERRESTRIAL | | | | | TERRESTRIAL | | | | |
|---|---|---|---|---|---|---|---|---|---|---|---|
| | | *Open Sun* | *Grass Sun* | *Bush Sun* | *Tree Sun* | *Other Sun* | *Open Shade* | *Grass Shade* | *Bush Shade* | *Tree Shade* | *Other Shade* |
| *Pogona minor* | 0.0 | 39.5 | 3.1 | 9.9 | 0.6 | 0.6 | 0.0 | 1.9 | 2.5 | 0.6 | 0.6 |
| *Varanus brevicauda* | 0.0 | 25.0 | 0.0 | 25.0 | 0.0 | 0.0 | 25.0 | 25.0 | 0.0 | 0.0 | 0.0 |
| *Varanus caudolineatus* | 0.0 | 16.7 | 0.0 | 0.0 | 0.0 | 0.0 | 0.0 | 0.0 | 0.0 | 8.3 | 0.0 |
| *Varanus eremius* | 0.0 | 21.3 | 25.9 | 6.5 | 0.9 | 0.9 | 0.9 | 38.0 | 3.7 | 1.9 | 0.0 |
| *Varanus gilleni* | 0.0 | 50.0 | 0.0 | 0.0 | 0.0 | 0.0 | 0.0 | 0.0 | 0.0 | 0.0 | 0.0 |
| *Varanus gouldi* | 0.0 | 75.0 | 3.2 | 2.6 | 1.3 | 0.0 | 4.5 | 2.6 | 0.6 | 3.2 | 0.6 |
| *Varanus tristis* | 0.0 | 3.8 | 3.2 | 0.6 | 1.9 | 1.3 | 1.3 | 3.2 | 0.6 | 1.9 | 1.3 |
| *Ctenotus ariadnae* | 0.0 | 11.8 | 38.2 | 0.0 | 0.0 | 0.0 | 0.0 | 50.0 | 0.0 | 0.0 | 0.0 |
| *Ctenotus atlas* | 0.0 | 4.2 | 25.0 | 0.0 | 0.0 | 0.0 | 0.0 | 70.8 | 0.0 | 0.0 | 0.0 |
| *Ctenotus brooksi* | 0.0 | 35.1 | 14.3 | 35.1 | 0.0 | 0.0 | 0.0 | 8.4 | 7.1 | 0.0 | 0.0 |
| *Ctenotus calurus* | 0.0 | 36.9 | 41.8 | 0.0 | 0.0 | 0.0 | 2.2 | 19.1 | 0.0 | 0.0 | 0.0 |
| *Ctenotus colletti* | 0.0 | 23.1 | 26.9 | 0.0 | 0.0 | 0.0 | 0.0 | 34.6 | 7.7 | 7.7 | 0.0 |
| *Ctenotus dux* | 0.0 | 14.3 | 36.3 | 8.5 | 0.0 | 0.9 | 2.7 | 35.7 | 1.2 | 0.0 | 0.3 |
| *Ctenotus grandis* | 0.0 | 15.5 | 10.7 | 0.0 | 0.0 | 0.0 | 0.0 | 73.8 | 0.0 | 0.0 | 0.0 |
| *Ctenotus helenae* | 0.0 | 4.5 | 12.4 | 0.0 | 5.3 | 0.0 | 0.0 | 68.0 | 2.3 | 6.8 | 0.0 |
| *Ctenotus leae* | 0.0 | 33.3 | 10.4 | 18.8 | 0.0 | 0.0 | 0.0 | 10.4 | 27.1 | 0.0 | 0.0 |
| *Ctenotus leonhardii* | 0.0 | 22.3 | 3.9 | 17.5 | 2.4 | 0.0 | 1.9 | 26.2 | 16.5 | 9.2 | 0.0 |
| *Ctenotus pantherinus* | 0.0 | 3.1 | 20.1 | 1.2 | 0.0 | 0.0 | 0.0 | 73.6 | 1.2 | 0.8 | 0.0 |
| *Ctenotus piankai* | 0.0 | 10.0 | 15.0 | 0.0 | 0.0 | 2.5 | 0.0 | 70.0 | 0.0 | 0.0 | 2.5 |
| *Ctenotus quattuordecimlineatus* | 0.0 | 8.1 | 18.2 | 0.7 | 0.1 | 0.3 | 0.5 | 69.1 | 0.8 | 1.9 | 0.3 |
| *Ctenotus schomburgkii* | 0.0 | 31.5 | 26.3 | 4.0 | 7.2 | 0.8 | 5.2 | 20.7 | 3.2 | 1.2 | 0.0 |
| *Cryptoblepharus plagiocephalus* | 0.0 | 2.7 | 0.9 | 0.0 | 0.9 | 4.9 | 0.9 | 0.0 | 0.0 | 1.2 | 1.3 |
| *Egernia depressa* | 0.0 | 4.0 | 0.0 | 0.0 | 0.0 | 0.0 | 0.0 | 0.0 | 0.0 | 0.0 | 0.0 |
| *Egernia kintorei* | 0.0 | 0.0 | 0.0 | 0.0 | 0.0 | 0.0 | 50.0 | 50.0 | 0.0 | 0.0 | 0.0 |
| *Egernia inornata* | 0.0 | 18.8 | 18.8 | 0.0 | 0.0 | 14.6 | 14.6 | 18.8 | 4.2 | 4.2 | 6.3 |
| *Egernia striata* | 0.0 | 11.7 | 6.9 | 1.7 | 0.0 | 13.5 | 17.7 | 20.2 | 5.5 | 0.0 | 22.8 |
| *Lerista bipes* | 90.2 | 0.0 | 0.0 | 0.6 | 0.6 | 0.0 | 0.0 | 0.0 | 0.6 | 3.0 | 4.9 |
| *Lerista desertorum* | 62.5 | 0.0 | 0.0 | 0.0 | 0.0 | 0.0 | 0.0 | 12.5 | 12.5 | 12.5 | 0.0 |
| *Lerista muelleri* | 90.0 | 0.0 | 0.0 | 0.0 | 0.0 | 0.0 | 0.0 | 0.0 | 0.0 | 10.0 | 0.0 |
| *Menetia greyi* | 0.0 | 15.2 | 39.4 | 0.0 | 4.5 | 3.0 | 4.5 | 16.7 | 9.1 | 4.5 | 3.0 |
| *Morethia butleri* | 0.0 | 0.0 | 4.5 | 0.0 | 22.7 | 0.0 | 0.0 | 4.5 | 0.0 | 50.0 | 0.0 |
| *Omolepida branchialis* | 0.0 | 0.0 | 0.0 | 16.7 | 0.0 | 0.0 | 0.0 | 83.3 | 0.0 | 0.0 | 0.0 |
| *Eremiascincus richardsoni* | 14.3 | 0.0 | 0.0 | 0.0 | 0.0 | 0.0 | 14.3 | 0.0 | 0.0 | 0.0 | 71.4 |
| *Tiliqua multifasciata* | 0.0 | 50.0 | 0.0 | 0.0 | 0.0 | 0.0 | 16.7 | 0.0 | 33.3 | 0.0 | 0.0 |
| *Delma fraseri* | 0.0 | 6.7 | 0.0 | 0.0 | 0.0 | 0.0 | 0.0 | 93.3 | 0.0 | 0.0 | 0.0 |
| *Lialis burtonis* | 0.0 | 18.2 | 18.2 | 0.0 | 0.0 | 0.0 | 9.1 | 54.5 | 0.0 | 0.0 | 0.0 |
| *Pygopus nigriceps* | 0.0 | 0.0 | 0.0 | 0.0 | 0.0 | 0.0 | 81.3 | 12.5 | 6.3 | 0.0 | 0.0 |
| *Diplodactylus ciliaris* | 0.0 | 0.0 | 0.0 | 0.0 | 0.0 | 0.0 | 11.1 | 4.3 | 10.3 | 4.3 | 1.7 |
| *Diplodactylus conspicillatus* | 0.0 | 0.0 | 0.0 | 0.0 | 0.0 | 0.0 | 58.1 | 31.2 | 4.3 | 0.0 | 5.4 |
| *Diplodactylus damaeus* | 0.0 | 0.0 | 0.0 | 0.0 | 0.0 | 0.0 | 41.9 | 19.4 | 33.9 | 4.8 | 0.0 |
| *Diplodactylus elderi* | 0.0 | 0.0 | 0.0 | 0.0 | 0.0 | 0.0 | 5.1 | 87.2 | 0.0 | 7.7 | 0.0 |
| *Diplodactylus pulcher* | 0.0 | 0.0 | 0.0 | 0.0 | 0.0 | 0.0 | 45.5 | 0.0 | 0.0 | 9.1 | 9.1 |
| *Diplodactylus stenodactylus* | 0.0 | 0.0 | 0.0 | 0.0 | 0.0 | 0.0 | 100.0 | 0.0 | 0.0 | 0.0 | 0.0 |
| *Diplodactylus strophurus* | 0.0 | 0.0 | 0.0 | 0.0 | 0.0 | 0.0 | 11.7 | 3.3 | 3.3 | 0.0 | 3.3 |
| *Gehyra variegata* | 0.0 | 0.0 | 0.0 | 0.0 | 0.0 | 0.0 | 3.3 | 1.0 | 2.1 | 4.8 | 2.4 |
| *Heteronotia binoei* | 0.0 | 0.0 | 0.0 | 0.0 | 0.0 | 0.0 | 40.9 | 4.5 | 9.1 | 18.2 | 22.7 |
| *Nephrurus laevissimus* | 0.0 | 0.0 | 0.0 | 0.0 | 0.0 | 0.0 | 65.4 | 13.5 | 18.8 | 0.8 | 1.5 |
| *Nephrurus levis* | 0.0 | 0.0 | 0.0 | 0.0 | 0.0 | 0.0 | 54.2 | 25.0 | 12.5 | 8.3 | 0.0 |
| *Nephrurus vertebralis* | 0.0 | 0.0 | 0.0 | 0.0 | 0.0 | 0.0 | 84.6 | 7.7 | 7.7 | 0.0 | 0.0 |
| *Rhynchoedura ornata* | 0.0 | 0.0 | 0.0 | 0.0 | 0.0 | 0.0 | 84.5 | 9.4 | 1.5 | 3.7 | 0.8 |
| Total number of lizards | 50 | 1003 | 968 | 231 | 41 | 29 | 832 | 1125 | 334 | 141 | 83 |

| Arboreal | | | | | |
|---|---|---|---|---|---|
| *Low Sun* | *Low Shade* | *High Sun* | *High Shade* | *Sample Size, N* | *Microhabitat Niche Breadth* |
| 13.6 | 3.7 | 16.0 | 7.4 | 81 | 4.57 |
| 0.0 | 0.0 | 0.0 | 0.0 | 2 | 4.00 |
| 0.0 | 0.0 | 37.5 | 37.5 | 12 | 3.16 |
| 0.0 | 0.0 | 0.0 | 0.0 | 54 | 3.80 |
| 0.0 | 0.0 | 0.0 | 50.0 | 2 | 2.00 |
| 3.8 | 1.3 | 1.3 | 0.0 | 78 | 1.75 |
| 0.0 | 9.0 | 6.4 | 65.4 | 78 | 2.25 |
| 0.0 | 0.0 | 0.0 | 0.0 | 17 | 2.44 |
| 0.0 | 0.0 | 0.0 | 0.0 | 24 | 1.77 |
| 0.0 | 0.0 | 0.0 | 0.0 | 77 | 3.59 |
| 0.0 | 0.0 | 0.0 | 0.0 | 202 | 2.87 |
| 0.0 | 0.0 | 0.0 | 0.0 | 13 | 3.89 |
| 0.0 | 0.0 | 0.0 | 0.0 | 164 | 3.48 |
| 0.0 | 0.0 | 0.0 | 0.0 | 103 | 1.72 |
| 0.0 | 0.0 | 0.4 | 0.4 | 133 | 2.05 |
| 0.0 | 0.0 | 0.0 | 0.0 | 24 | 4.14 |
| 0.0 | 0.0 | 0.0 | 0.0 | 103 | 5.34 |
| 0.0 | 0.0 | 0.0 | 0.0 | 127 | 1.71 |
| 0.0 | 0.0 | 0.0 | 0.0 | 20 | 1.91 |
| 0.0 | 0.0 | 0.0 | 0.0 | 366 | 1.93 |
| 0.0 | 0.0 | 0.0 | 0.0 | 126 | 4.51 |
| 27.9 | 7.8 | 32.3 | 19.3 | 113 | 4.36 |
| 16.0 | 8.0 | 20.0 | 52.0 | 25 | 2.91 |
| 0.0 | 0.0 | 0.0 | 0.0 | 2 | 2.00 |
| 0.0 | 0.0 | 0.0 | 0.0 | 24 | 6.44 |
| 0.0 | 0.0 | 0.0 | 0.0 | 72 | 6.10 |
| 0.0 | 0.0 | 0.0 | 0.0 | 41 | 1.22 |
| 0.0 | 0.0 | 0.0 | 0.0 | 4 | 2.29 |
| 0.0 | 0.0 | 0.0 | 0.0 | 10 | 1.22 |
| 0.0 | 0.0 | 0.0 | 0.0 | 17 | 4.50 |
| 9.1 | 0.0 | 9.1 | 0.0 | 11 | 3.10 |
| 0.0 | 0.0 | 0.0 | 0.0 | 6 | 1.38 |
| 0.0 | 0.0 | 0.0 | 0.0 | 7 | 1.81 |
| 0.0 | 0.0 | 0.0 | 0.0 | 3 | 2.57 |
| 0.0 | 0.0 | 0.0 | 0.0 | 15 | 1.14 |
| 0.0 | 0.0 | 0.0 | 0.0 | 11 | 2.69 |
| 0.0 | 0.0 | 0.0 | 0.0 | 16 | 1.47 |
| 0.0 | 34.2 | 0.0 | 34.2 | 117 | 3.84 |
| 0.0 | 1.1 | 0.0 | 0.0 | 93 | 2.28 |
| 0.0 | 0.0 | 0.0 | 0.0 | 62 | 3.03 |
| 0.0 | 0.0 | 0.0 | 0.0 | 39 | 1.30 |
| 0.0 | 31.8 | 0.0 | 4.5 | 22 | 3.06 |
| 0.0 | 0.0 | 0.0 | 0.0 | 3 | 1.00 |
| 1.7 | 61.7 | 0.0 | 1.5 | 60 | 2.38 |
| 0.0 | 40.2 | 0.0 | 46.2 | 909 | 2.64 |
| 0.0 | 0.0 | 0.0 | 4.5 | 22 | 3.78 |
| 0.0 | 0.0 | 0.0 | 0.0 | 375 | 2.08 |
| 0.0 | 0.0 | 0.0 | 0.0 | 24 | 2.64 |
| 0.0 | 0.0 | 0.0 | 0.0 | 13 | 1.37 |
| 0.0 | 0.0 | 0.0 | 0.0 | 361 | 1.38 |
| 90 | 487 | 127 | 587 | 6128 | 3.02 (Mean) |

# Appendix D
# Body Temperature Statistics

Appendix D.1. Body temperature statistics for 12 species of North American desert lizards.

| | Body Temperature | | | Air Temperature | | | | |
|---|---|---|---|---|---|---|---|---|
| Species | $\overline{X}$ | *SD* | *N* | $\overline{X}$ | *SD* | *N* | *Slope* | *Intercept* |
| *Cnemidophorus tigris* | 39.5 | 1.8 | 1848 | 27.1 | 3.8 | 1783 | 0.41 | 28.3 |
| *Uta stansburiana* | 35.3 | 2.4 | 778 | 26.1 | 4.6 | 592 | 0.02 | 34.7 |
| *Crotaphytus wislizeni* | 37.4 | 2.2 | 183 | 26.0 | 3.4 | 133 | 0.17 | 32.8 |
| *Phrynosoma platyrhinos* | 34.9 | 3.2 | 211 | 26.1 | 4.5 | 109 | 0.40 | 24.0 |
| *Callisaurus draconoides* | 39.1 | 2.6 | 352 | 29.1 | 4.1 | 201 | 0.41 | 26.3 |
| *Sceloporus magister* | 34.8 | 1.6 | 92 | 31.3 | 3.1 | 29 | 0.34 | 24.6 |
| *Urosaurus graciosus* | 36.2 | 1.5 | 32 | 30.9 | 3.1 | 34 | 0.33 | 25.9 |
| *Urosaurus ornatus* | 35.6 | 2.1 | 117 | 27.5 | 2.9 | 117 | 0.18 | 30.8 |
| *ipsosaurus dorsalis* | 40.0 | 2.3 | 60 | 31.2 | 5.3 | 41 | 0.29 | 31.1 |
| *ma scoparia* | 37.3 | 2.2 | 30 | 28.3 | 3.8 | 31 | 0.07 | 35.3 |
| *antusia vigilis* | 29.3 | 2.0 | 20 | 26.5 | 2.2 | 24 | 0.71 | 10.7 |
| *oleonyx variegata* | 28.4 | 3.4 | 35 | 30.2 | 2.3 | 28 | 0.76 | 7.1 |

Appendix D.2. Body temperature statistics for 19 species of Kalahari desert lizards.

| | Body Temperature | | | Air Temperature | | | | |
|---|---|---|---|---|---|---|---|---|
| Species | $\overline{X}$ | *SD* | *N* | $\overline{X}$ | *SD* | *N* | *Slope* | *Intercept* |
| *Agama hispida* | 36.2 | 2.8 | 215 | 28.6 | 4.5 | 264 | 0.28 | 27.7 |
| *Eremias lineo-ocellata* | 36.9 | 2.3 | 649 | 28.9 | 5.1 | 886 | 0.27 | 29.1 |
| *Eremias lugubris* | 37.7 | 2.5 | 176 | 29.1 | 3.1 | 221 | 0.37 | 27.0 |
| *Eremias namaquensis* | 37.8 | 2.1 | 151 | 30.1 | 3.5 | 195 | 0.19 | 32.0 |
| *Ichnotropis squamulosa* | 36.3 | 1.8 | 92 | 31.3 | 2.5 | 110 | 0.21 | 29.6 |
| *Meroles suborbitalis* | 35.5 | 2.2 | 480 | 26.5 | 5.5 | 663 | 0.22 | 29.9 |
| *Nucras intertexta* | 38.9 | — | 3 | 34.0 | — | 4 | — | — |
| *Nucras tessellata* | 39.3 | 3.1 | 43 | 31.6 | 4.2 | 84 | 0.49 | 24.1 |
| *Mabuya occidentalis* | 36.0 | 2.7 | 136 | 30.0 | 3.5 | 164 | 0.43 | 23.1 |
| *Mabuya spilogaster* | 34.5 | 2.8 | 271 | 28.8 | 4.9 | 309 | 0.28 | 26.7 |
| *Mabuya striata* | 34.1 | 2.8 | 281 | 28.1 | 4.6 | 369 | 0.29 | 25.9 |
| *Mabuya variegata* | 33.6 | 2.5 | 20 | 28.2 | 6.6 | 71 | 0.26 | 27.2 |
| *Chondrodactylus angulifer* | 25.4 | 3.5 | 349 | 26.1 | 3.3 | 375 | 0.96 | 0.3 |
| *Colopus wahlbergi* | 25.3 | 3.3 | 58 | 25.9 | 3.3 | 87 | 0.91 | 2.2 |
| *Lygodactylus capensis* | 36.0 | — | 1 | 26.9 | — | 14 | — | — |
| *Pachydactylus bibroni* | 27.9 | 3.0 | 58 | 26.1 | 2.8 | 69 | 0.71 | 9.7 |
| *Pachydactylus capensis* | 25.3 | 3.7 | 44 | 25.2 | 4.0 | 54 | 0.83 | 4.8 |
| *Pachydactylus rugosus* | 25.3 | 3.9 | 11 | 25.6 | 3.8 | 17 | 1.03 | −1.0 |
| *Ptenopus garrulus* | 27.6 | 3.5 | 148 | 26.1 | 3.6 | 246 | 0.80 | 7.2 |

Appendix D.3. Body temperature statistics for Australian desert lizards.

| Species | Body Temperature | | | Air Temperature | | | | |
|---|---|---|---|---|---|---|---|---|
| | $\overline{X}$ | SD | N | $\overline{X}$ | SD | N | Slope | Intercept |
| *Caimanops amphiboluroides* | 36.6 | 2.6 | 13 | 32.5 | 2.8 | 13 | 0.67 | 14.7 |
| *Ctenophorus clayi* | 36.7 | 2.6 | 23 | 29.4 | 4.1 | 23 | 0.21 | 30.6 |
| *Ctenophorus fordi* | 36.9 | 2.2 | 110 | 26.3 | 4.7 | 110 | 0.23 | 30.9 |
| *Ctenophorus inermis* | 36.1 | 3.9 | 155 | 30.2 | 5.5 | 155 | 0.52 | 20.5 |
| *Ctenophorus isolepis* | 37.8 | 2.3 | 1261 | 27.6 | 5.1 | 1261 | 0.23 | 31.6 |
| *Ctenophorus reticulatus* | 34.4 | 3.5 | 30 | 29.1 | 4.7 | 30 | 0.49 | 20.2 |
| *Ctenophorus scutulatus* | 38.9 | 1.9 | 81 | 31.8 | 4.1 | 81 | 0.27 | 30.3 |
| *Diporiphora winneckei* | 33.7 | 3.3 | 32 | 25.5 | 6.2 | 32 | 0.35 | 24.8 |
| *Gemmatophora longirostris* | 34.1 | 3.9 | 68 | 25.4 | 6.3 | 68 | 0.45 | 22.6 |
| *Moloch horridus* | 32.6 | 4.1 | 190 | 27.1 | 5.2 | 190 | 0.54 | 18.1 |
| *Pogona minor* | 34.6 | 3.8 | 81 | 27.6 | 4.9 | 81 | 0.40 | 23.5 |
| *Varanus caudolineatus* | 37.8 | 3.5 | 10 | 33.6 | 3.5 | 10 | 0.71 | 14.0 |
| *Varanus eremius* | 37.5 | 3.0 | 53 | 28.3 | 6.0 | 53 | 0.24 | 30.8 |
| *Varanus gouldi* | 37.5 | 3.5 | 67 | 30.0 | 4.4 | 67 | 0.49 | 22.8 |
| *Varanus tristis* | 31.8 | 6.6 | 57 | 28.5 | 5.6 | 57 | 0.89 | 6.5 |
| *Ctenotus ariadnae* | 36.3 | 1.8 | 15 | 30.5 | 3.3 | 15 | 0.48 | 21.7 |
| *Ctenotus atlas* | 35.0 | 3.1 | 22 | 29.9 | 4.7 | 22 | 0.22 | 28.4 |
| *Ctenotus brooksi* | 31.1 | 2.9 | 87 | 21.5 | 5.0 | 87 | 0.35 | 23.6 |
| *Ctenotus calurus* | 36.0 | 2.1 | 182 | 27.1 | 3.3 | 182 | 0.24 | 29.4 |
| *Ctenotus colletti* | 36.6 | 1.4 | 10 | 29.9 | 3.7 | 10 | 0.15 | 32.2 |
| *Ctenotus dux* | 32.2 | 3.1 | 163 | 22.7 | 4.1 | 163 | 0.25 | 26.6 |
| *Ctenotus grandis* | 35.2 | 3.1 | 82 | 28.1 | 5.1 | 82 | 0.38 | 24.6 |
| *Ctenotus helenae* | 33.4 | 3.7 | 115 | 27.1 | 5.5 | 115 | 0.48 | 20.3 |
| *Ctenotus leae* | 38.0 | 1.8 | 21 | 28.5 | 4.2 | 21 | 0.11 | 34.9 |
| *Ctenotus leonhardii* | 38.1 | 1.9 | 94 | 30.2 | 3.7 | 94 | 0.14 | 33.8 |
| *Ctenotus pantherinus* | 33.9 | 3.2 | 113 | 28.0 | 6.3 | 113 | 0.39 | 22.9 |
| *Ctenotus piankai* | 35.8 | 3.8 | 20 | 29.9 | 4.2 | 20 | 0.66 | 16.0 |
| *Ctenotus quattuordecimlineatus* | 35.8 | 3.0 | 298 | 29.1 | 4.3 | 298 | 0.42 | 23.6 |
| *Ctenotus schomburgkii* | 33.4 | 3.1 | 115 | 26.9 | 4.4 | 115 | 0.38 | 23.1 |
| *Cryptoblepharus plagiocephalus* | 32.9 | 3.5 | 91 | 25.6 | 5.6 | 91 | 0.13 | 29.6 |
| *Egernia depressa* | 34.0 | 3.3 | 28 | 31.0 | 4.0 | 28 | 0.67 | 13.2 |
| *Egernia inornata* | 30.1 | 3.5 | 103 | 26.5 | 5.3 | 103 | 0.52 | 16.5 |
| *Egernia striata* | 30.9 | 4.0 | 145 | 28.1 | 4.9 | 145 | 0.60 | 14.0 |
| *Lerista bipes* | 31.2 | 4.5 | 4 | 27.4 | 4.4 | 4 | 0.77 | 10.1 |
| *Menetia greyi* | 33.0 | 3.2 | 7 | 27.7 | 5.8 | 7 | 0.45 | 20.6 |
| *Morethia butleri* | 33.6 | 3.8 | 11 | 29.2 | 5.8 | 11 | 0.46 | 20.1 |
| *Omolepida branchialis* | 34.7 | 3.0 | 3 | 30.3 | 3.7 | 3 | — | — |
| *Eremiascincus richardsoni* | 26.2 | 2.0 | 6 | 24.9 | 5.1 | 6 | 0.37 | 17.0 |
| *Tiliqua multifasciata* | 34.3 | 1.3 | 2 | 26.5 | 7.8 | 2 | — | — |
| *Delma fraseri* | 31.7 | 2.3 | 8 | 29.1 | 3.6 | 8 | 0.56 | 15.4 |
| *Lialis burtonis* | 29.0 | 5.0 | 11 | 25.5 | 5.8 | 11 | 0.79 | 8.8 |
| *Pygopus nigriceps* | 24.7 | 2.3 | 11 | 23.7 | 1.9 | 11 | 0.99 | 1.4 |
| *Diplodactylus ciliaris* | 26.0 | 5.0 | 113 | 26.3 | 4.9 | 113 | 0.99 | −0.1 |
| *Diplodactylus conspicillatus* | 28.0 | 3.5 | 83 | 27.7 | 3.7 | 83 | 0.85 | 4.5 |
| *Diplodactylus damaeus* | 27.3 | 3.6 | 55 | 26.6 | 3.8 | 55 | 0.86 | 4.4 |
| *Diplodactylus elderi* | 28.2 | 7.0 | 16 | 27.6 | 6.6 | 16 | 1.04 | −0.5 |
| *Diplodactylus pulcher* | 27.7 | 3.7 | 24 | 27.5 | 3.5 | 24 | 0.75 | 7.0 |
| *Diplodactylus strophurus* | 25.3 | 4.9 | 63 | 25.7 | 4.9 | 63 | 0.98 | 0.3 |
| *Gehyra variegata* | 27.4 | 4.1 | 840 | 26.5 | 4.2 | 840 | 0.90 | 3.6 |
| *Heteronotia binoei* | 27.5 | 3.5 | 24 | 25.7 | 4.8 | 24 | 0.65 | 10.7 |
| *Nephrurus laevissimus* | 24.9 | 4.5 | 360 | 25.1 | 4.3 | 360 | 0.97 | 0.5 |
| *Nephrurus levis* | 23.1 | 3.4 | 32 | 22.5 | 3.9 | 32 | 0.76 | 6.0 |
| *Nephrurus vertebralis* | 24.1 | 3.5 | 14 | 24.1 | 3.4 | 14 | 0.99 | 0.3 |
| *Rhynchoedura ornata* | 27.7 | 3.1 | 330 | 26.9 | 3.3 | 330 | 0.79 | 6.5 |

# Appendix E
## Diet Summaries and Food Niche Breadths

Appendix E.1. Percentage utilization of various food resource states among North American desert lizard species, with various totals and dietary niche breadths.

| | FOOD RESOURCE CATEGORY | | | | | | | | | | |
|---|---|---|---|---|---|---|---|---|---|---|---|
| Species | Ce* | Sp | Sc | So | A | W | G | B | M | N | Co |
| *Cnemidophorus tigris* | | 1.9 | 1.3 | 2.1 | 0.4 | 0.4 | 11.1 | 4.8 | 1.0 | 0.3 | 17.2 |
| *Uta stansburiana* | | 3.9 | — | 0.5 | 10.3 | 1.3 | 18.1 | 1.5 | 0.9 | 0.4 | 23.5 |
| *Phrynosoma platyrhinos* | | 0.7 | — | 2.1 | 58.9 | 0.3 | 0.5 | 2.8 | 0.1 | — | 23.7 |
| *Crotaphytus wislizeni* | | 0.5 | 0.1 | — | 0.5 | 0.7 | 22.7 | 0.1 | 0.4 | — | 11.1 |
| *Callisaurus draconoides* | | 2.2 | 0.1 | 0.1 | 3.4 | 4.3 | 26.0 | 2.9 | 1.2 | 0.4 | 23.5 |
| *Sceloporus magister* | | 0.4 | — | 1.9 | 23.6 | 2.7 | 1.4 | 0.8 | 0.5 | 0.1 | 51.5 |
| *Urosaurus graciosus* | | 0.1 | — | — | 16.4 | 8.9 | 7.5 | — | 7.5 | — | 25.4 |
| *Dipsosaurus dorsalis* | | — | — | — | 0.2 | 0.2 | — | — | — | — | 0.8 |
| *Uma scoparia* | | 1.9 | 1.9 | 1.0 | 51.9 | 0.2 | 1.4 | 3.0 | — | 1.4 | 11.5 |
| *Coleonyx variegata* | | 4.3 | 1.4 | 0.7 | — | 2.9 | 19.4 | 4.3 | — | — | 24.5 |
| *Xantusia vigilis* | | 7.7 | — | — | 23.0 | — | — | 7.7 | — | — | 38.3 |
| Total volume of prey, cc. (in stomachs of all species) | | 50.5 | 23.2 | 45.7 | 307.4 | 27.6 | 363.8 | 100.0 | 24.5 | 7.0 | 587.2 |

* Centipedes are not eaten by these North American lizards.

| FOOD RESOURCE CATEGORY | | | | | | | | | Total Stomach Volume (cc.) | Food Niche Breadth | Total Number of Lizards |
|---|---|---|---|---|---|---|---|---|---|---|---|
| I | H | D | Lp | E | Lv | U | V | P | | | |
| 30.0 | 0.6 | 0.4 | 3.8 | 0.4 | 18.1 | 2.6 | 3.6 | 0.1 | 1582.2 | 5.82 | 1975 |
| 14.7 | 5.8 | 2.3 | 1.0 | 0.1 | 7.4 | 6.5 | 0.2 | 1.6 | 228.9 | 7.33 | 944 |
| 0.4 | 0.7 | — | — | 0.1 | 8.1 | 0.7 | 0.1 | 0.8 | 342.8 | 2.43 | 140 |
| — | 0.1 | 2.0 | 1.4 | 0.7 | 2.2 | 1.7 | 52.3 | 3.4 | 334.6 | 2.94 | 162 |
| 3.9 | 1.3 | 2.4 | 0.4 | 0.3 | 12.5 | 4.5 | 4.7 | 5.7 | 245.8 | 6.55 | 441 |
| — | 1.0 | 0.5 | 0.1 | — | 7.9 | 1.7 | 0.4 | 5.6 | 138.2 | 3.02 | 88 |
| 11.9 | 3.0 | 16.4 | — | — | — | 3.0 | — | — | 6.7 | 6.53 | 43 |
| — | — | 0.5 | — | — | 0.1 | 0.9 | — | 97.3 | 220.0 | 1.06 | 63 |
| 0.7 | 0.7 | 0.3 | — | 0.2 | 3.2 | 5.2 | 2.4 | 13.3 | 59.2 | 3.26 | 32 |
| 29.5 | 0.7 | — | — | — | 7.9 | 4.3 | — | — | 13.9 | 5.07 | 49 |
| 7.7 | — | 7.7 | 7.7 | — | 0.4 | — | — | — | 1.3 | 4.37 | 24 |
| 525.0 | 30.6 | 27.4 | 67.9 | 10.0 | 384.2 | 83.5 | 245.9 | 262.4 | — | 4.40 (Mean down) | — |

## Key to Food Resource Categories

Ce Centipedes
Sp Spiders
Sc Scorpions
So Solpugids (absent from Australia)
A Ants
W Wasps and other non-ant hymenopterans
G Grasshopper and crickets
B Roaches (Blattids)
M Mantids and phasmids
N Adult Neuroptera (ant lions)
Co Beetles (Coleoptera)
I Termites (Isoptera)
H Bugs (Hemiptera and Homoptera)
D Flies (Diptera)
Lp Butterflies and moths (Lepidoptera)
E Insect eggs and pupae
Lv All insect larvae
U Miscellaneous arthropods, including unidentified items
V All vertebrate material, including sloughed lizard skins
P Plant materials (floral and vegetative)

Appendix E.2. Percentage utilization of various food resource states among Kalahari lizard species, with various totals and estimates of dietary niche breadths.

| Species | Food Resource Category | | | | | | | | | | |
|---|---|---|---|---|---|---|---|---|---|---|---|
| | Ce* | Sp | Sc | So | A | W | G | B | M | N | Co |
| *Agama hispida* | — | 0.9 | 0.9 | 1.8 | 48.7 | 0.6 | 0.6 | 0.01 | — | 0.01 | 23.0 |
| *Eremias lineo-ocellata* | — | 8.6 | 0.5 | 1.3 | 2.3 | 2.0 | 11.3 | 0.3 | — | 0.2 | 15.2 |
| *Eremias lugubris* | — | 0.8 | 0.8 | 0.1 | 0.5 | 0.1 | 0.9 | 0.3 | — | 0.01 | 0.9 |
| *Eremias namaquensis* | — | 7.8 | 0.7 | 1.1 | 3.9 | 0.7 | 4.7 | 0.1 | 0.01 | 0.01 | 2.3 |
| *Ichnotropis squamulosa* | 0.1 | 3.9 | 0.01 | 0.7 | 0.1 | 0.01 | 11.5 | 0.2 | 0.01 | 0.01 | 0.01 |
| *Meroles suborbitalis* | 0.1 | 4.6 | 0.3 | 2.3 | 5.0 | 1.4 | 6.6 | 0.01 | 0.1 | — | 14.6 |
| *Nucras intertexta* | — | 1.0 | 1.0 | — | 0.4 | — | 36.0 | 2.2 | — | — | 13.1 |
| *Nucras tessellata* | — | 9.5 | 53.2 | 0.01 | 0.2 | 0.01 | 17.2 | 0.01 | 0.01 | 0.01 | 1.0 |
| *Mabuya occidentalis* | 0.1 | 2.2 | 3.1 | 5.1 | 3.9 | 2.0 | 11.3 | 0.01 | 0.1 | 0.1 | 36.9 |
| *Mabuya spilogaster* | 0.1 | 3.0 | 0.01 | 0.9 | 2.2 | 1.6 | 9.1 | 1.6 | 0.01 | 0.01 | 23.3 |
| *Mabuya striata* | 0.1 | 4.8 | 1.0 | 0.3 | 11.2 | 1.2 | 3.5 | 0.8 | 0.4 | 0.01 | 29.0 |
| *Mabuya variegata* | — | 5.9 | — | 0.3 | 1.2 | 0.7 | 15.4 | 0.9 | 2.9 | 1.0 | 10.2 |
| *Typhlosaurus gariepensis* | — | 0.8 | 0.01 | 0.01 | 0.4 | 0.01 | 0.01 | 0.01 | 0.01 | 0.01 | — |
| *Typhlosaurus lineatus* | — | 0.3 | — | 0.1 | 0.2 | 0.01 | 0.01 | 0.01 | 0.01 | 0.01 | 2.4 |
| *Chondrodactylus angulifer* | — | 1.9 | 10.5 | 2.4 | 0.5 | 0.01 | 12.4 | 0.4 | 0.2 | 0.01 | 10.9 |
| *Colopus wahlbergi* | — | 5.5 | 0.2 | 1.3 | 3.5 | 0.2 | 7.1 | 0.9 | 0.9 | 0.4 | — |
| *Lygodactylus capensis* | — | 11.4 | 0.01 | 0.01 | 29.3 | 0.01 | 9.0 | 0.01 | 0.01 | 0.01 | 6.5 |
| *Pachydactylus bibroni* | 0.9 | 1.6 | 1.0 | 1.2 | 3.7 | 0.1 | 4.4 | 1.4 | 0.01 | 0.01 | 16.9 |
| *Pachydactylus capensis* | 0.3 | 3.5 | 3.4 | 2.9 | 2.7 | 0.01 | 8.8 | 0.9 | 0.01 | 1.1 | 10.2 |
| *Pachydactylus rugosus* | — | 3.5 | 0.01 | 1.4 | 0.6 | 0.01 | 4.0 | 14.7 | 1.4 | 0.01 | 11.8 |
| *Ptenopus garrulus* | 0.2 | 2.5 | 0.01 | 0.6 | 12.6 | 0.3 | 2.1 | 0.3 | 0.01 | — | 7.2 |
| Total volume of prey, cc. (in stomachs of all species) | 0.8 | 35.7 | 32.7 | 17.6 | 155.3 | 8.7 | 70.0 | 4.2 | 1.2 | 0.5 | 186.7 |

Appendix E.3. Percentage utilization of various food resource states among Australian desert lizard species, with various totals and dietary niche breadths.

| Species | Food Resource Category | | | | | | | | | |
|---|---|---|---|---|---|---|---|---|---|---|
| | Ce* | Sp | Sc | So | A | W | G | B | M | N |
| *Caimanops amphiboluroides* | 3.3 | — | — | — | 1.4 | — | — | — | — | — |
| *Ctenophorus clayi* | 1.8 | 0.9 | — | — | 10.0 | — | 40.8 | 3.1 | — | 5.8 |
| *Ctenophorus fordi* | 0.9 | 3.0 | 0.1 | — | 48.9 | 9.1 | 5.9 | — | — | 0.2 |
| *Ctenophorus inermis* | 3.6 | 1.0 | 0.2 | — | 16.8 | 6.6 | 6.6 | 0.5 | 0.3 | — |
| *Ctenophorus isolepis* | 1.0 | 1.4 | 0.1 | — | 53.7 | 4.0 | 8.7 | 1.0 | 0.3 | 0.1 |
| *Ctenophorus reticulatus* | 2.4 | 1.5 | — | — | 28.1 | 0.1 | 5.6 | — | 0.01 | — |
| *Ctenophorus scutulatus* | 5.7 | 6.2 | 0.2 | — | 31.0 | 4.3 | 7.1 | 0.6 | 0.2 | — |
| *Diporiphora winneckei* | — | 2.7 | — | — | 5.4 | 19.9 | 1.9 | — | 5.2 | 1.9 |
| *Gemmatophora longirostris* | — | 0.9 | — | — | 6.6 | 27.1 | 11.0 | 0.8 | 6.1 | 1.4 |
| *Moloch horridus* | — | — | — | — | 99.1 | — | — | — | — | — |
| *Pogona minor* | 2.1 | 0.5 | — | — | 4.2 | 1.5 | 29.9 | 0.8 | 0.2 | 0.1 |

| FOOD RESOURCE CATEGORY | | | | | | | | | *Total Stomach Volume (cc.)* | *Food Niche Breadth* | *Total Number of Lizards* |
|---|---|---|---|---|---|---|---|---|---|---|---|
| I | H | D | Lp | E | Lv | U | V | P | | | |
| 14.4 | 0.4 | 0.3 | 0.3 | — | 1.9 | 1.7 | 1.0 | 3.4 | 248.7 | 3.20 | 368 |
| 43.5 | 4.1 | 1.4 | 0.3 | 0.01 | 4.5 | 3.6 | 0.5 | 0.4 | 111.0 | 4.18 | 1135 |
| 92.8 | 0.5 | 0.1 | 0.2 | 0.01 | 1.1 | 0.8 | — | — | 59.0 | 1.16 | 238 |
| 66.7 | 0.7 | 0.2 | 0.01 | 0.01 | 4.8 | 5.5 | 0.8 | 0.1 | 25.2 | 2.17 | 218 |
| 79.2 | — | 0.01 | 0.01 | 0.01 | 3.1 | 1.2 | 0.01 | 0.01 | 23.7 | 1.55 | 112 |
| 51.7 | 2.2 | 1.8 | 0.3 | 0.01 | 5.9 | 2.5 | 0.3 | 0.3 | 92.9 | 3.30 | 780 |
| 15.6 | — | — | 14.4 | — | — | 15.3 | — | — | 2.3 | 4.63 | 6 |
| 4.6 | 0.01 | 0.1 | 0.9 | 0.01 | 6.5 | 6.7 | 0.01 | 0.01 | 10.6 | 3.01 | 79 |
| 23.2 | 1.8 | 0.7 | 0.9 | 0.01 | 3.3 | 3.4 | 1.0 | 1.0 | 49.6 | 4.72 | 211 |
| 30.1 | 3.2 | 1.5 | 10.0 | 0.01 | 8.7 | 3.1 | 1.5 | 0.1 | 49.5 | 5.71 | 367 |
| 22.5 | 2.7 | 1.2 | 5.6 | — | 7.4 | 5.1 | 2.6 | 0.5 | 117.6 | 6.10 | 554 |
| 30.4 | 12.3 | 4.4 | 1.4 | 0.01 | 3.8 | 7.7 | 0.01 | 1.6 | 3.5 | 6.42 | 105 |
| 96.8 | 0.01 | 0.01 | 0.01 | 0.1 | 1.9 | 0.01 | 0.01 | 0.01 | 9.5 | 1.07 | 125 |
| 93.3 | 0.01 | 0.01 | 0.01 | 0.1 | 3.4 | 0.1 | 0.01 | 0.01 | 41.5 | 1.15 | 454 |
| 45.8 | — | 0.01 | 0.2 | 0.01 | 0.4 | 5.1 | 8.9 | 0.3 | 187.1 | 3.85 | 407 |
| 63.6 | 0.8 | 0.01 | 1.5 | 0.01 | 2.7 | 8.2 | 2.1 | 1.0 | 11.7 | 2.37 | 122 |
| 10.2 | 8.3 | 4.2 | 0.01 | 0.01 | 8.7 | 9.6 | 3.0 | 0.01 | 0.8 | 6.78 | 27 |
| 51.5 | 0.3 | 0.01 | 2.2 | 0.01 | 8.3 | 5.5 | 2.0 | 0.01 | 45.9 | 3.25 | 150 |
| 32.1 | 1.8 | 0.01 | 5.4 | 0.6 | 8.2 | 12.3 | 6.1 | 0.1 | 8.8 | 6.49 | 84 |
| 8.1 | 5.1 | 1.4 | 22.2 | 1.4 | 5.5 | 13.7 | 4.3 | 0.6 | 1.7 | 8.22 | 26 |
| 62.4 | 0.8 | — | 0.2 | 0.01 | 0.9 | 5.7 | 0.2 | 4.3 | 45.9 | 2.40 | 440 |
| 473.3 | 15.2 | 6.8 | 16.3 | 0.2 | 41.2 | 40.0 | 26.3 | 13.2 | — | 3.89 (Mean down) | — |

| FOOD RESOURCE CATEGORY | | | | | | | | | | *Total Stomach Volume (cc.)* | *Food Niche Breadth* | *Total Number of Lizards* |
|---|---|---|---|---|---|---|---|---|---|---|---|---|
| Co | I | H | D | Lp | E | Lv | U | V | P | | | |
| — | 87.1 | — | — | — | — | — | 0.9 | — | 7.3 | 7.69 | 1.31 | 13 |
| 8.9 | 1.3 | 6.7 | 0.4 | — | 0.9 | 0.4 | 7.5 | — | 11.5 | 2.25 | 4.70 | 26 |
| 2.2 | 0.2 | 14.3 | 2.0 | 2.7 | — | 2.7 | 5.5 | 0.4 | 2.0 | 13.18 | 3.60 | 113 |
| 11.3 | 15.7 | 4.2 | 0.1 | — | 0.01 | 6.7 | 0.8 | 0.1 | 25.3 | 162.63 | 6.84 | 186 |
| 3.7 | 13.5 | 4.9 | 1.6 | 0.4 | 0.01 | 2.7 | 1.7 | 0.3 | 0.9 | 249.44 | 3.11 | 1357 |
| 1.9 | 28.8 | 1.9 | — | — | — | 2.4 | 0.1 | — | 27.3 | 34.01 | 4.14 | 46 |
| 3.8 | 26.3 | 1.6 | 0.9 | 0.2 | — | 6.1 | 3.2 | 0.4 | 2.0 | 46.79 | 5.37 | 86 |
| 11.1 | — | 10.7 | 2.3 | — | — | 6.9 | 4.6 | — | 27.5 | 2.61 | 6.52 | 37 |
| 9.4 | 2.2 | 13.1 | 2.6 | — | — | 11.4 | 4.8 | — | 2.7 | 25.02 | 7.30 | 80 |
| 0.01 | 0.01 | — | — | 0.01 | — | — | 0.1 | — | 0.8 | 239.21 | 1.02 | 209 |
| 17.3 | 11.7 | 0.8 | 0.2 | — | 0.1 | 8.1 | 0.8 | 2.6 | 19.3 | 125.16 | 5.56 | 98 |

| SPECIES | FOOD RESOURCE CATEGORY | | | | | | | | | |
|---|---|---|---|---|---|---|---|---|---|---|
| | Ce* | Sp | Sc | So | A | W | G | B | M | N |
| *Varanus brevicauda* | — | — | — | — | — | — | 39.6 | 1.9 | — | — |
| *Varanus caudolineatus* | 4.7 | 11.9 | — | — | — | — | 13.4 | 17.4 | — | — |
| *Varanus eremius* | 0.5 | 0.1 | 2.8 | — | — | — | 17.6 | 3.2 | — | — |
| *Varanus gilleni* | — | — | 0.9 | — | — | — | 13.9 | — | — | — |
| *Varanus gouldi* | 4.7 | 5.0 | 5.2 | — | — | 0.1 | 6.4 | 3.2 | 0.2 | — |
| *Varanus tristis* | — | 0.5 | — | — | 0.1 | — | 10.6 | 1.5 | 0.5 | — |
| *Ctenotus ariadnae* | — | 1.9 | — | — | 1.4 | — | 2.8 | — | — | — |
| *Ctenotus atlas* | — | 0.4 | — | — | 0.4 | — | 17.3 | 3.8 | — | — |
| *Ctenotus brooksi* | — | 11.5 | — | — | 10.8 | 8.1 | 4.7 | 1.2 | — | 3.8 |
| *Ctenotus calurus* | — | 1.6 | — | — | 0.8 | 0.3 | 0.4 | 0.4 | 0.4 | 0.4 |
| *Ctenotus colletti* | — | 33.8 | — | — | — | — | 1.5 | 1.5 | — | — |
| *Ctenotus dux* | 1.1 | 4.2 | 2.4 | — | 5.5 | 0.3 | 18.4 | 1.2 | 1.3 | 0.4 |
| *Ctenotus grandis* | 3.3 | 0.3 | 0.4 | — | 1.4 | 0.4 | 0.01 | 1.3 | — | — |
| *Ctenotus helenae* | 0.3 | 1.6 | — | — | 1.1 | 0.8 | 7.3 | 6.4 | 0.4 | — |
| *Ctenotus leae* | — | 8.3 | — | — | 4.3 | 3.1 | 6.8 | — | 3.1 | — |
| *Ctenotus leonhardii* | 6.0 | 2.0 | 0.9 | — | 1.2 | 2.3 | 6.7 | 1.5 | 1.4 | — |
| *Ctenotus pantherinus* | 0.1 | 1.1 | — | — | 0.6 | 0.8 | 1.0 | 0.1 | — | — |
| *Ctenotus piankai* | — | 16.9 | — | — | 1.7 | — | 11.9 | — | 5.1 | — |
| *Ctenotus quattuordecimlineatus* | 0.2 | 2.1 | 0.01 | — | 1.5 | 0.7 | 7.4 | 1.0 | 1.8 | 0.01 |
| *Ctenotus schomburgkii* | — | 1.8 | — | — | 0.9 | 0.2 | 0.3 | 3.0 | — | — |
| *Cryptoblepharus plagiocephalus* | — | 18.0 | — | — | 6.3 | 0.7 | 9.2 | 5.6 | 0.7 | — |
| *Egernia depressa* | 2.8 | — | — | — | 0.8 | — | — | — | — | — |
| *Egernia kintorei* | — | — | — | — | 5.0 | — | — | 10.0 | — | — |
| *Egernia inornata* | 1.2 | 3.9 | 0.3 | — | 38.9 | 1.2 | 7.4 | 3.8 | 1.4 | — |
| *Egernia striata* | 0.7 | 0.9 | 0.01 | — | 10.0 | 0.3 | 0.7 | 2.4 | 0.01 | — |
| *Lerista bipes* | — | 0.8 | — | — | 1.7 | — | — | 2.1 | 1.7 | — |
| *Lerista desertorum* | — | — | — | — | — | — | — | — | — | — |
| *Lerista muelleri* | — | 6.3 | — | — | 15.6 | — | — | — | — | — |
| *Menetia greyi* | — | 30.0 | — | — | — | — | — | — | — | — |
| *Morethia butleri* | — | 8.5 | — | — | — | 22.7 | 21.3 | — | — | — |
| *Omolepida branchialis* | — | 0.4 | — | — | 0.4 | — | 16.3 | 3.6 | — | — |
| *Eremiascincus richardsoni* | — | — | — | — | 21.4 | 17.9 | — | — | — | — |
| *Tiliqua multifasciata* | 1.2 | — | — | — | 1.4 | — | 5.0 | — | 0.5 | — |
| *Delma fraseri* | — | 50.4 | — | — | — | — | 12.6 | 9.4 | — | — |
| *Lialis burtonis* | — | — | — | — | — | — | — | — | — | — |
| *Pygopus nigriceps* | — | 25.8 | 34.3 | — | 2.1 | — | — | — | — | — |
| *Diplodactylus ciliaris* | — | 10.4 | 0.2 | — | 0.1 | — | 33.7 | 9.8 | 7.6 | 0.6 |
| *Diplodactylus conspicillatus* | — | — | — | — | — | — | — | — | — | — |
| *Diplodactylus damaeus* | — | 11.0 | — | — | — | — | 24.6 | 1.7 | — | 1.0 |
| *Diplodactylus elderi* | — | 21.0 | — | — | 0.4 | 0.4 | 5.8 | 9.3 | — | — |
| *Diplodactylus pulcher* | — | — | — | — | — | — | — | — | — | — |
| *Diplodactylus stenodactylus* | — | 14.1 | — | — | 12.7 | — | — | — | — | — |
| *Diplodactylus strophurus* | — | 20.2 | — | — | — | — | 19.6 | 14.5 | 4.1 | — |
| *Gehyra variegata* | 0.8 | 6.9 | 0.3 | — | 0.4 | 0.4 | 8.9 | 9.5 | 1.2 | 0.2 |
| *Heteronotia binoei* | 2.7 | 25.3 | — | — | 4.3 | 0.5 | 10.8 | 9.1 | — | — |
| *Nephrurus laevissimus* | 6.9 | 11.8 | 8.6 | — | 0.1 | 0.01 | 21.2 | 20.3 | 4.3 | 0.2 |
| *Nephrurus levis* | 9.7 | 20.2 | 13.7 | — | — | — | 20.0 | 9.7 | — | — |
| *Nephrurus vertebralis* | 16.3 | 24.2 | 15.4 | — | — | — | 5.3 | — | — | — |
| *Rhynchoedura ornata* | — | 0.1 | — | — | 0.1 | — | — | — | — | — |
| Total volume of prey, cc. (in stomachs of all species) | 56.2 | 76.2 | 41.6 | — | 467.5 | 38.1 | 239.5 | 77.9 | 15.8 | 1.85 |

*Category 4 (Solpugids) is not represented in stomachs of Australian desert lizards.

| Co | I | H | D | Lp | E | Lv | U | V | P | Total Stomach Volume (*cc.*) | *Food Niche Breadth* | *Total Number of Lizards* |
|---|---|---|---|---|---|---|---|---|---|---|---|---|
| FOOD RESOURCE CATEGORY | | | | | | | | | | | | |
| 9.4 | — | — | — | — | — | 5.7 | 5.7 | 37.7 | — | 5.30 | 3.17 | 2 |
| — | — | 0.8 | — | 1.2 | — | 10.7 | 0.8 | 38.7 | 0.4 | 25.30 | 4.42 | 13 |
| — | — | — | — | — | — | 0.1 | 2.6 | 73.2 | — | 94.0 | 1.76 | 60 |
| — | — | — | — | — | — | — | — | 85.2 | — | 11.5 | 1.34 | 2 |
| 9.0 | — | — | — | 0.6 | — | 2.9 | 1.1 | 61.7 | — | 504.00 | 2.49 | 63 |
| 0.7 | — | 0.2 | — | — | — | 0.7 | 4.4 | 80.6 | — | 404.60 | 1.51 | 64 |
| 3.3 | 81.2 | 6.6 | — | 0.9 | — | — | 1.9 | — | — | 2.13 | 1.50 | 16 |
| 7.7 | 56.9 | 8.1 | — | 2.3 | — | 0.4 | 2.3 | 0.4 | — | 2.60 | 2.71 | 25 |
| 19.0 | 0.7 | 11.9 | 0.6 | — | 5.7 | 13.3 | 1.4 | 0.6 | 6.7 | 4.96 | 9.00 | 79 |
| 0.2 | 83.7 | 0.1 | — | — | 0.1 | 5.5 | 5.6 | 0.1 | 0.4 | 11.40 | 1.41 | 202 |
| — | 29.2 | 1.5 | 7.7 | — | — | 9.2 | 15.4 | — | — | 0.65 | 4.19 | 14 |
| 4.1 | 37.2 | 6.0 | 1.6 | 1.8 | 0.6 | 1.7 | 5.6 | 1.5 | 5.0 | 21.82 | 5.25 | 186 |
| 2.7 | 78.3 | 0.2 | 0.01 | 0.3 | 0.1 | 3.1 | 2.0 | 3.1 | 2.9 | 124.00 | 1.62 | 105 |
| 4.1 | 66.6 | 0.3 | — | 1.1 | 0.1 | 4.7 | 1.1 | 1.4 | 2.8 | 71.32 | 2.18 | 134 |
| 1.9 | — | 21.9 | 0.6 | — | — | 4.0 | 5.7 | — | 40.2 | 3.24 | 4.34 | 22 |
| 1.3 | 31.6 | 4.0 | — | 1.6 | — | 25.9 | 1.7 | 1.2 | 10.8 | 27.81 | 5.25 | 116 |
| 0.8 | 84.6 | 0.5 | 0.2 | 0.3 | — | 1.5 | 2.0 | 0.8 | 5.6 | 46.22 | 1.39 | 134 |
| 1.7 | — | 34.7 | — | 2.5 | — | 5.1 | 11.0 | — | 9.3 | 1.18 | 5.24 | 22 |
| 3.8 | 57.9 | 7.4 | 0.2 | 3.2 | 0.2 | 4.9 | 2.9 | 0.6 | 4.0 | 65.27 | 2.82 | 388 |
| 1.9 | 71.8 | 1.5 | 0.1 | — | 0.8 | 14.0 | 2.9 | — | 0.6 | 9.26 | 1.86 | 134 |
| 14.4 | 11.6 | 10.2 | 5.3 | 1.1 | 0.7 | 7.4 | 6.7 | 2.1 | — | 2.84 | 9.43 | 104 |
| 0.1 | 85.5 | — | — | — | — | 10.1 | 0.6 | — | 0.1 | 18.13 | 1.35 | 32 |
| — | — | — | — | 2.5 | — | — | — | — | 82.5 | 4.00 | 1.44 | 1 |
| 11.6 | 7.8 | 1.0 | 1.0 | 0.5 | — | 5.4 | 1.6 | 3.2 | 9.9 | 30.36 | 5.16 | 124 |
| 7.6 | 67.5 | 0.3 | — | 0.1 | — | 1.1 | 0.7 | 4.6 | 3.2 | 112.36 | 2.11 | 190 |
| 17.6 | 55.2 | — | 2.5 | — | — | 2.5 | 14.2 | 1.7 | — | 1.20 | 2.79 | 92 |
| 8.3 | 45.8 | 25.0 | — | — | 16.7 | — | 4.2 | — | — | 0.24 | 3.24 | 8 |
| 3.1 | 12.5 | — | 18.8 | — | — | — | 25.0 | 18.8 | — | 0.16 | 5.63 | 10 |
| 1.7 | 36.7 | 16.7 | — | — | — | — | 15.0 | — | — | 0.30 | 3.64 | 20 |
| — | 39.7 | — | — | — | — | — | 7.8 | — | — | 0.71 | 3.73 | 16 |
| 12.7 | 53.6 | 7.6 | — | 2.2 | — | 0.4 | 2.5 | 0.4 | — | 2.76 | 2.95 | 6 |
| 35.7 | 17.9 | — | — | — | — | — | 7.1 | — | — | 0.28 | 4.13 | 8 |
| 7.8 | 3.7 | 4.1 | — | — | — | 0.6 | 0.9 | — | 74.7 | 8.03 | 1.75 | 3 |
| — | 18.9 | 3.9 | — | — | — | 1.6 | 3.1 | — | — | 1.27 | 3.15 | 18 |
| — | — | — | — | — | — | — | — | 100.0 | — | 1.50 | 1.00 | 12 |
| — | — | — | — | — | — | — | 37.8 | — | — | 4.66 | 3.06 | 18 |
| 12.3 | — | 4.2 | 0.2 | 2.9 | — | 14.8 | 2.5 | 0.01 | 0.7 | 25.30 | 5.55 | 115 |
| — | 99.5 | — | — | — | — | — | 0.4 | — | 0.1 | 7.90 | 1.01 | 91 |
| 24.9 | 5.6 | 5.3 | 7.6 | 9.3 | — | 3.3 | 5.6 | — | — | 3.01 | 6.26 | 63 |
| 5.8 | 40.0 | 7.4 | 1.7 | 0.8 | — | 2.7 | 4.3 | — | 0.4 | 2.58 | 4.39 | 45 |
| — | 98.0 | — | — | — | — | — | — | — | 2.0 | 1.01 | 1.04 | 25 |
| — | — | 56.3 | — | — | — | — | 16.9 | — | — | 0.36 | 2.62 | 3 |
| 12.9 | — | 4.8 | 0.4 | 5.8 | — | 12.4 | 5.2 | — | — | 8.46 | 7.01 | 67 |
| 12.6 | 35.5 | 5.8 | 0.5 | 4.8 | 0.4 | 2.9 | 3.8 | 4.3 | 0.7 | 98.93 | 5.76 | 871 |
| 5.4 | 7.0 | 4.3 | — | 7.5 | — | 4.8 | 7.0 | 10.2 | 1.1 | 1.86 | 8.37 | 44 |
| 14.2 | 0.4 | 0.6 | — | 0.2 | 0.1 | 2.5 | 3.8 | 2.2 | 2.6 | 78.01 | 7.29 | 368 |
| 13.2 | — | — | — | — | — | 11.0 | 2.5 | — | — | 9.79 | 6.73 | 35 |
| 3.5 | — | — | — | — | — | — | 0.4 | 35.0 | — | 4.92 | 4.25 | 12 |
| — | 98.6 | — | — | — | — | — | 0.9 | — | 0.3 | 11.56 | 1.03 | 367 |
| 159.7 | 515.5 | 48.4 | 7.8 | 16.2 | 1.7 | 92.7 | 58.4 | 753.9 | 118.1 | — | 3.81 (Mean down) | — |

# Appendix F
# Reproductive Statistics

Appendix F. Statistics on reproductive tactics of 65 species of desert lizards in ten families from three continents.

| Family and Species | Clutch/Litter Size | | | Reproductive Effort | | | Expenditure per Progeny | | |
|---|---|---|---|---|---|---|---|---|---|
| | $\overline{X}$ | SD | N | $\overline{X}$ | SD | N | $\overline{X}$ | SD | N |
| AGAMIDAE | | | | | | | | | |
| *Agama hispida* | 13.4 | 3.1 | 45 | 30.7 | 8.4 | 22 | 2.21 | 0.66 | 312 |
| *Ctenophorus caudicinctus* | 5.4 | 2.0 | 18 | 16.6 | 3.0 | 10 | 3.02 | 0.68 | 41 |
| *Ctenophorus clayi* | 1.9 | 0.4 | 7 | 12.9 | 6.8 | 6 | 6.58 | 3.06 | 12 |
| *Ctenophorus fordi* | 2.4 | 0.5 | 15 | 11.8 | 2.6 | 5 | 6.02 | 1.43 | 10 |
| *Ctenophorus inermis* | 4.1 | 1.3 | 27 | 15.1 | 3.9 | 8 | 3.21 | 0.98 | 33 |
| *Ctenophorus isolepis* | 3.0 | 1.0 | 211 | 12.9 | 4.4 | 76 | 4.23 | 1.27 | 178 |
| *Ctenophorus reticulatus* | 4.0 | 1.2 | 7 | 19.2 | 0.8 | 3 | 5.34 | 0.95 | 10 |
| *Ctenophorus scutulatus* | 6.8 | 1.9 | 6 | 11.8 | 3.4 | 2 | 1.83 | 0.15 | 9 |
| *Gemmatophora longirostris* | 3.9 | 0.9 | 7 | 15.1 | 1.8 | 4 | 3.55 | 1.32 | 17 |
| *Moloch horridus* | 6.8 | 1.6 | 84 | 14.9 | 3.8 | 34 | 2.27 | 0.81 | 202 |
| *Pogona minor* | 7.6 | 2.3 | 73 | 19.5 | 4.4 | 27 | 2.33 | 0.92 | 219 |
| EUBLEPHARIDAE | | | | | | | | | |
| *Coleonyx variegata* | 2.0 | 0.0 | 43 | 18.3 | 1.7 | 14 | 9.35 | 3.17 | 26 |
| GEKKONIDAE | | | | | | | | | |
| *Chondrodactylus angulifer* | 1.9 | 0.3 | 92 | 12.2 | 3.5 | 22 | 6.02 | 1.62 | 41 |
| *Colopus wahlbergi* | 2.0 | 0.0 | 17 | 8.2 | 1.9 | 2 | 4.34 | 0.72 | 3 |
| *Diplodactylus ciliaris* | 2.0 | 0.0 | 31 | 10.9 | 2.9 | 20 | 5.42 | 1.45 | 40 |
| *Diplodactylus conspicillatus* | 2.0 | 0.0 | 18 | 19.6 | 1.4 | 10 | 10.24 | 1.95 | 18 |
| *Diplodactylus damaeus* | 2.0 | 0.0 | 11 | 13.4 | 3.3 | 4 | 6.67 | 3.04 | 8 |
| *Diplodactylus elderi* | 2.0 | 0.0 | 6 | 13.2 | 1.6 | 5 | 6.60 | 1.92 | 10 |
| *Diplodactylus pulcher* | 1.8 | 0.4 | 12 | 14.7 | 4.9 | 5 | 9.01 | 4.16 | 7 |
| *Diplodactylus strophurus* | 2.0 | 0.0 | 30 | 12.6 | 2.7 | 18 | 6.25 | 1.39 | 34 |
| *Gehyra variegata* | 1.0 | 0.0 | 152 | 5.3 | 1.6 | 92 | 5.34 | 1.57 | 92 |
| *Heteronotia binoei* | 2.0 | 0.0 | 8 | 9.9 | 2.0 | 3 | 5.24 | 1.45 | 6 |
| *Nephrurus laevissimus* | 1.96 | 0.21 | 66 | 16.1 | 4.0 | 29 | 7.78 | 2.46 | 58 |
| *Nephrurus levis* | 2.0 | 0.0 | 5 | 16.1 | 0.3 | 2 | 7.93 | 0.61 | 2 |
| *Pachydactylus bibroni* | 2.0 | 0.0 | 16 | 10.6 | 1.8 | 8 | 5.16 | 0.85 | 14 |
| *Pachydactylus capensis* | 1.93 | 0.26 | 14 | 11.1 | 2.5 | 3 | 5.48 | 1.47 | 6 |
| *Ptenopus garrulus* | 1.0 | 0.0 | 14 | 8.8 | 0.0 | 1 | 8.80 | 0.0 | 1 |
| *Rhynchoedura ornata* | 1.97 | 0.23 | 73 | 16.8 | 4.65 | 35 | 8.53 | 2.69 | 50 |

Appendix F. (*continued*)

| Family and Species | Clutch/Litter Size | | | Reproductive Effort | | | Expenditure per Progeny | | |
|---|---|---|---|---|---|---|---|---|---|
| | $\overline{X}$ | *SD* | *N* | $\overline{X}$ | *SD* | *N* | $\overline{X}$ | *SD* | *N* |
| IGUANIDAE | | | | | | | | | |
| *Callisaurus draconoides* | 4.3 | 1.3 | 73 | 15.2 | 3.2 | 14 | 4.11 | 0.93 | 17 |
| *Crotaphytus wislizeni* | 5.3 | 1.9 | 94 | 15.6 | 5.6 | 11 | 4.14 | 1.00 | 36 |
| *Dipsosaurus dorsalis* | 4.7 | 1.8 | 87 | 16.1 | 2.7 | 11 | 3.22 | 0.44 | 12 |
| *Phrynosoma platyrhinos* | 7.8 | 2.2 | 128 | 21.0 | 6.9 | 36 | 3.07 | 0.88 | 31 |
| *Sceloporus magister* | 8.4 | — | 14 | 20.7 | — | 4 | 2.02 | 0.19 | 6 |
| *Uma scoparia* | 3.1 | 0.6 | 9 | 13.3 | 4.2 | 4 | 3.71 | 0.50 | 14 |
| *Urosaurus graciosus* | 3.8 | 1.1 | 5 | 18.2 | 0.7 | 3 | 3.72 | 0.40 | 10 |
| *Uta stansburiana* | 3.79 | 1.32 | 353 | 22.9 | 8.0 | 108 | 5.33 | 1.68 | 331 |
| LACERTIDAE | | | | | | | | | |
| *Eremias lineo-ocellata* | 6.94 | 2.0 | 123 | 18.7 | 5.4 | 42 | 2.75 | 0.73 | 261 |
| *Eremias lugubris* | 3.86 | 0.98 | 50 | 14.6 | 3.1 | 22 | 3.95 | 0.64 | 83 |
| *Eremias namaquensis* | 3.9 | 1.5 | 29 | 16.5 | 6.1 | 5 | 4.40 | 0.92 | 9 |
| *Ichnotropis squamulosa* | 4.0 | 0.0 | 2 | 15.0 | 0.0 | 1 | 4.00 | 0.42 | 4 |
| *Meroles suborbitalis* | 4.23 | 0.87 | 108 | 21.9 | 5.0 | 23 | 5.36 | 1.02 | 88 |
| *Nucras tessellata* | 3.25 | 0.66 | 8 | 15.8 | 0.0 | 1 | 5.50 | 0.24 | 3 |
| PYGOPODIDAE | | | | | | | | | |
| *Delma fraseri* | 2.0 | 0.0 | 1 | 13.2 | 0.0 | 1 | 6.57 | 0.24 | 2 |
| *Lialis burtoni* | 2.0 | 0.0 | 2 | 11.4 | 0.0 | 1 | 5.68 | 0.23 | 2 |
| *Pygopus nigriceps* | 1.5 | 0.71 | 2 | 23.5 | 0.0 | 1 | 11.76 | 1.19 | 2 |
| SCINCIDAE | | | | | | | | | |
| *Cryptoblepharus plagiocephalus* | 2.0 | 0.0 | 8 | 22.3 | 6.1 | 7 | 11.16 | 3.04 | 14 |
| *Ctenotus colletti* | 4.0 | 0.0 | 1 | 18.8 | 0.0 | 1 | 9.38 | 0.0 | 2 |
| *Ctenotus pantherinus* | 5.8 | 1.6 | 17 | 16.5 | 3.9 | 10 | 2.59 | 0.64 | 65 |
| *Ctenotus quattuordecimlineatus* | 3.6 | 1.1 | 7 | 3.9 | 0.7 | 3 | 1.20 | 0.49 | 6 |
| *Egernia inornata* | 2.1 | 0.7 | 32 | 13.4 | 5.1 | 21 | 6.66 | 2.26 | 42 |
| *Egernia striata* | 2.7 | 1.0 | 19 | 9.5 | 3.7 | 18 | 3.74 | 1.60 | 36 |
| *Lerista bipes* | 2.0 | 0.0 | 3 | 18.2 | 5.1 | 3 | 9.12 | 2.28 | 6 |
| *Mabuya occidentalis* | 6.7 | 1.6 | 32 | 16.0 | 3.8 | 12 | 2.53 | 0.78 | 72 |
| *Mabuya spilogaster* | 4.4 | 1.3 | 74 | 19.9 | 5.2 | 32 | 4.73 | 1.76 | 70 |
| *Mabuya striata* | 5.4 | 1.6 | 74 | 23.9 | 5.5 | 48 | 4.05 | 1.23 | 79 |
| *Mabuya variegata* | 2.0 | | 13 | 11.0 | 6.2 | 6 | 5.39 | 1.75 | 10 |
| *Omolepida branchialis* | 3.0 | 0.0 | 1 | 18.6 | 0.0 | 1 | 6.19 | 1.65 | 3 |
| *Eremiascincus richardsoni* | 5.0 | 0.0 | 1 | 11.4 | 0.0 | 1 | 2.28 | 0.35 | 5 |
| *Typhlosaurus gariepensis* | 1.0 | 0.0 | 11 | 17.3 | 3.7 | 7 | 17.26 | 3.73 | 7 |
| *Typhlosaurus lineatus* | 1.5 | 0.1 | 90 | 18.9 | 7.9 | 23 | 11.49 | 2.48 | 36 |
| TEIDAE | | | | | | | | | |
| *Cnemidophorus tigris* | 2.5 | 0.9 | 293 | 11.3 | 3.7 | 64 | 4.53 | 1.52 | 75 |
| VARANIDAE | | | | | | | | | |
| *Varanus eremius* | 3.6 | 1.1 | 8 | 16.1 | 2.3 | 2 | 4.59 | 1.15 | 7 |
| *Varanus gouldi* | 6.4 | 1.4 | 11 | 13.9 | 2.8 | 2 | 1.70 | 0.46 | 16 |
| *Varanus tristis* | 10.2 | 2.8 | 19 | 16.2 | 2.9 | 11 | 1.49 | 0.31 | 120 |
| XANTUSIDAE | | | | | | | | | |
| *Xantusia vigilis* | 1.84 | — | 164 | 10.9 | 3.5 | 16 | 6.04 | 1.74 | 12 |

# Appendix G
# Anatomical Statistics

Appendix G.1. Estimates of morphometrics for 12 species of North American desert lizards (in millimeters).

| Species/Statistic | Snout-Vent Length | Tail Length[a] | Head Length | Head Width | Head Depth | Jaw Length |
|---|---|---|---|---|---|---|
| *Cnemidophorus tigris* | | | | | | |
| $\overline{X}$ | 79.4 | 185.4 | 18.6 | 9.9 | 8.6 | 21.1 |
| *SD* | 9.6 | 36.4 | 2.4 | 1.7 | 1.6 | 2.6 |
| *N* | 147 | 915 | 146 | 145 | 145 | 146 |
| *Uta stansburiana* | | | | | | |
| $\overline{X}$ | 42.2 | 63.3 | 9.8 | 6.9 | 4.5 | 11.3 |
| *SD* | 9.7 | 15.5 | 1.9 | 1.6 | 1.0 | 2.1 |
| *N* | 312 | 668 | 301 | 304 | 303 | 301 |
| *Crotaphytus wislizeni* | | | | | | |
| $\overline{X}$ | 92.6 | 183.0 | 22.9 | 17.4 | 12.0 | 25.7 |
| *SD* | 14.1 | 30.2 | 3.4 | 2.6 | 1.6 | 3.7 |
| *N* | 154 | 167 | 154 | 154 | 154 | 154 |
| *Phrynosoma platyrhinos* | | | | | | |
| $\overline{X}$ | 71.8 | 41.0 | 13.6 | 14.4 | 11.6 | 11.9 |
| *SD* | 12.5 | 10.0 | 1.7 | 1.8 | 1.6 | 1.7 |
| *N* | 164 | 169 | 157 | 153 | 148 | 153 |
| *Callisaurus draconoides* | | | | | | |
| $\overline{X}$ | 73.4 | 89.9 | 14.2 | 11.8 | 8.8 | 15.4 |
| *SD* | 10.9 | 17.5 | 1.6 | 1.3 | 1.2 | 1.8 |
| *N* | 276 | 283 | 261 | 259 | 252 | 250 |
| *Sceloporus magister* | | | | | | |
| $\overline{X}$ | 86.3 | 105.6 | 19.2 | 16.7 | 11.2 | 21.5 |
| *SD* | 21.7 | 32.7 | 4.2 | 4.8 | 3.1 | 5.4 |
| *N* | 85 | 34 | 84 | 84 | 84 | 84 |
| *Urosaurus graciosus* | | | | | | |
| $\overline{X}$ | 48.7 | 103.4 | 11.3 | 8.2 | 6.2 | 12.9 |
| *SD* | 6.7 | 18.3 | 1.3 | 1.1 | 0.9 | 1.3 |
| *N* | 35 | 35 | 33 | 33 | 33 | 33 |
| *Dipsosaurus dorsalis* | | | | | | |
| $\overline{X}$ | 107.0 | 181.6 | 18.5 | 16.2 | 12.1 | 21.6 |
| *SD* | 23.8 | 44.2 | 3.4 | 3.2 | 2.4 | 4.1 |
| *N* | 62 | 62 | 63 | 63 | 60 | 60 |

| FOREFOOT LENGTH | FOREARM LENGTH | HIND-FOOT LENGTH | HIND-LEG LENGTH |
|---|---|---|---|
| 13.6 | 29.5 | 31.8 | 59.1 |
| 1.6 | 3.4 | 3.4 | 6.8 |
| 147 | 147 | 147 | 147 |
| | | | |
| 7.9 | 18.2 | 14.6 | 31.0 |
| 1.6 | 3.7 | 2.7 | 6.5 |
| 316 | 316 | 316 | 314 |
| | | | |
| 17.3 | 37.3 | 35.9 | 73.5 |
| 2.3 | 4.8 | 4.7 | 10.1 |
| 154 | 154 | 154 | 154 |
| | | | |
| 13.2 | 34.4 | 18.0 | 44.8 |
| 2.2 | 6.0 | 2.9 | 8.1 |
| 165 | 165 | 165 | 165 |
| | | | |
| 16.2 | 39.9 | 32.9 | 69.0 |
| 2.1 | 5.2 | 4.1 | 8.9 |
| 278 | 277 | 278 | 279 |
| | | | |
| 17.8 | 40.0 | 26.9 | 58.5 |
| 4.9 | 8.9 | 5.1 | 13.6 |
| 85 | 85 | 85 | 85 |
| | | | |
| 8.0 | 19.8 | 15.9 | 31.5 |
| 1.0 | 3.3 | 2.0 | 4.7 |
| 35 | 34 | 35 | 35 |
| | | | |
| 17.3 | 39.7 | 35.8 | 76.4 |
| 3.5 | 8.3 | 6.4 | 14.3 |
| 64 | 64 | 64 | 64 |

Appendix G.1. (*continued*)

| Species/Statistic | Snout-Vent Length | Tail Length[a] | Head Length | Head Width | Head Depth | Jaw Length |
|---|---|---|---|---|---|---|
| *Uma scoparia* | | | | | | |
| $\overline{X}$ | 81.6 | 81.6 | 16.8 | 13.4 | 9.6 | 19.0 |
| *SD* | 15.9 | 16.5 | 2.5 | 2.1 | 1.7 | 2.9 |
| *N* | 44 | 43 | 44 | 44 | 44 | 43 |
| *Xantusia vigilis* | | | | | | |
| $\overline{X}$ | 42.2 | 51.1 | 7.9 | 5.1 | 3.7 | 9.6 |
| *SD* | 4.5 | 8.7 | 0.6 | 0.5 | 0.3 | 0.7 |
| *N* | 27 | 27 | 27 | 27 | 27 | 27 |
| *Coleonyx variegatus* | | | | | | |
| $\overline{X}$ | 50.6 | 41.7 | 10.8 | 8.1 | 5.0 | 11.8 |
| *SD* | 9.9 | 11.0 | 1.8 | 1.4 | 0.9 | 2.0 |
| *N* | 180 | 45 | 180 | 180 | 180 | 180 |
| *Heloderma suspectum* | | | | | | |
| $\overline{X}$ | 254.7 | 191[b] | 45.4 | 38.7 | 23.3 | 50.7 |
| *SD* | 62.4 | — | 9.3 | 10.1 | 6.1 | 10.9 |
| *N* | 40 | — | 45 | 44 | 44 | 44 |

[a]Tail lengths estimated from field measurements on fresh material, adjusted for snout-vent length and preservation shrinkage.
[b]Not actually measured, but estimated from the literature.

Appendix G.2. Morphometric statistics for 21 species of Kalahari desert lizards (in millimeters).

| Species | Snout-Vent Length | Tail Length | Head Length | Head Width | Head Depth | Jaw Length |
|---|---|---|---|---|---|---|
| *Agama hispida* | | | | | | |
| Mean | 66.6 | 89.3 | 16.0 | 15.8 | 10.8 | 19.0 |
| SD | 25.7 | 38.0 | 5.0 | 4.9 | 3.2 | 6.4 |
| *Chamaeleo dilepis* | | | | | | |
| Mean | 94.5 | 96.0 | 24.1 | 16.1 | 17.3 | 25.0 |
| SD | 13.4 | 8.5 | 3.1 | 2.6 | 2.8 | 3.3 |
| *Eremias lineo-ocellata* | | | | | | |
| Mean | 47.6 | 99.8 | 11.5 | 7.2 | 5.2 | 13.0 |
| SD | 12.0 | 24.6 | 2.5 | 1.6 | 1.3 | 2.7 |
| *Eremias lugubris* | | | | | | |
| Mean | 50.1 | 117.1 | 12.2 | 8.3 | 6.4 | 13.4 |
| SD | 10.4 | 27.5 | 2.3 | 5.5 | 1.4 | 2.6 |
| *Eremias namaquensis* | | | | | | |
| Mean | 48.1 | 107.0 | 10.9 | 6.9 | 5.0 | 11.9 |
| SD | 7.1 | 22.5 | 1.4 | 0.9 | 0.9 | 1.5 |
| *Ichnotropis squamulosa* | | | | | | |
| Mean | 51.3 | 109.5 | 12.3 | 8.4 | 6.6 | 13.5 |
| SD | 9.0 | 22.1 | 1.9 | 1.3 | 1.1 | 2.1 |
| *Meroles suborbitalis* | | | | | | |
| Mean | 51.3 | 111.7 | 12.2 | 7.8 | 5.7 | 13.2 |
| SD | 9.3 | 23.6 | 1.9 | 1.4 | 2.7 | 2.0 |

| FOREFOOT LENGTH | FOREARM LENGTH | HIND-FOOT LENGTH | HIND-LEG LENGTH |
|---|---|---|---|
| 15.3 | 37.1 | 27.3 | 60.6 |
| 2.2 | 6.5 | 3.6 | 9.9 |
| 44 | 44 | 44 | 44 |
| 4.5 | 11.5 | 6.9 | 14.7 |
| 0.6 | 1.3 | 0.9 | 1.3 |
| 27 | 27 | 27 | 27 |
| 5.6 | 16.5 | 8.6 | 21.4 |
| 1.1 | 3.0 | 1.7 | 3.8 |
| 180 | 180 | 179 | 180 |
| 30.3 | 73.8 | 30.4 | 79.1 |
| 5.8 | 14.9 | 6.6 | 17.5 |
| 45 | 46 | 46 | 46 |

| FOREFOOT LENGTH | FOREARM LENGTH | HIND-FOOT LENGTH | HIND-LEG LENGTH |
|---|---|---|---|
| 12.7 | 31.3 | 18.4 | 45.9 |
| 3.7 | 10.1 | 5.3 | 14.2 |
| 15.0 | 46.0 | 16.8 | 43.0 |
| 1.4 | 7.1 | 0.4 | 9.9 |
| 8.6 | 19.1 | 18.3 | 34.6 |
| 2.0 | 4.6 | 4.0 | 8.6 |
| 8.8 | 18.6 | 20.2 | 37.2 |
| 1.4 | 3.5 | 3.5 | 7.2 |
| 8.3 | 17.1 | 17.7 | 32.8 |
| 1.2 | 2.4 | 2.4 | 4.9 |
| 9.3 | 19.0 | 18.0 | 33.5 |
| 0.5 | 3.0 | 2.8 | 5.6 |
| 8.8 | 19.0 | 18.7 | 36.6 |
| 2.4 | 3.0 | 2.7 | 5.6 |

| Species | Snout-vent length | Tail length | Head length | Head width | Head depth | Jaw length |
|---|---|---|---|---|---|---|
| *Nucras tessellata* | | | | | | |
| Mean | 56.8 | 126.0 | 11.7 | 7.9 | 5.9 | 14.2 |
| SD | 13.2 | 29.6 | 2.2 | 1.9 | 1.5 | 2.8 |
| *Mabuya occidentalis* | | | | | | |
| Mean | 74.7 | 110.9 | 14.8 | 11.5 | 8.2 | 18.2 |
| SD | 12.7 | 20.4 | 2.1 | 1.5 | 1.1 | 2.6 |
| *Mabuya spilogaster* | | | | | | |
| Mean | 61.9 | 86.0 | 13.2 | 9.8 | 6.8 | 15.2 |
| SD | 11.3 | 16.1 | 1.9 | 1.6 | 1.3 | 2.4 |
| *Mabuya striata* | | | | | | |
| Mean | 70.1 | 99.0 | 15.5 | 11.2 | 8.0 | 18.2 |
| SD | 16.4 | 19.9 | 2.9 | 2.5 | 1.9 | 3.5 |
| *Mabuya variegata* | | | | | | |
| Mean | 36.6 | 54.5 | 7.6 | 5.0 | 3.9 | 9.0 |
| SD | 5.8 | 8.8 | 0.8 | 0.7 | 0.6 | 1.2 |
| *Typhlosaurus gariepensis* | | | | | | |
| Mean | 119.4 | 21.5 | 3.8 | 2.8 | 2.2* | 4.0* |
| SD | 5.8 | 3.3 | 0.2 | 0.1 | — | — |
| *Typhlosaurus lineatus* | | | | | | |
| Mean | 136.1 | 20.6 | 4.5 | 3.3 | 2.6* | 4.8* |
| SD | 9.6 | 6.2 | 0.3 | 0.2 | — | — |
| *Chondrodactylus angulifer* | | | | | | |
| Mean | 75.5 | 57.0 | 17.9 | 16.4 | 11.5 | 18.9 |
| SD | 16.5 | 11.7 | 3.7 | 4.1 | 2.9 | 4.0 |
| *Colopus wahlbergi* | | | | | | |
| Mean | 47.8 | 44.8 | 10.8 | 8.3 | 5.8 | 11.5 |
| SD | 6.1 | 6.1 | 1.4 | 1.1 | 0.8 | 1.4 |
| *Lygodactylus capensis* | | | | | | |
| Mean | 30.9 | 31.8 | 7.6 | 5.4 | 3.7 | 8.4 |
| SD | 5.9 | 7.8 | 1.2 | 0.9 | 0.7 | 1.1 |
| *Pachydactylus bibroni* | | | | | | |
| Mean | 65.1 | 62.4 | 16.2 | 14.7 | 8.5 | 17.1 |
| SD | 15.1 | 15.5 | 3.4 | 3.6 | 2.3 | 3.6 |
| *Pachydactylus capensis* | | | | | | |
| Mean | 51.6 | 53.4 | 12.0 | 9.2 | 5.8 | 12.5 |
| SD | 10.3 | 11.0 | 2.2 | 1.8 | 1.4 | 2.4 |
| *Pachydactylus rugosus* | | | | | | |
| Mean | 52.0 | 39.3 | 12.9 | 10.2 | 6.3 | 14.0 |
| SD | 6.0 | 7.4 | 1.3 | 1.2 | 0.8 | 1.4 |
| *Ptenopus garrulus* | | | | | | |
| Mean | 43.0 | 32.0 | 9.2 | 7.6 | 5.3 | 9.5 |
| SD | 7.1 | 6.0 | 1.3 | 1.3 | 0.9 | 1.4 |

*Not actually measured, but estimated from other head proportions.

Note: Tail lengths measured on fresh material, adjusted for preservation shrinkage and to snout-vent length.

| FOREFOOT LENGTH | FOREARM LENGTH | HIND-FOOT LENGTH | HIND-LEG LENGTH |
|---|---|---|---|
| 8.7 | 16.5 | 15.9 | 30.3 |
| 1.8 | 3.8 | 3.5 | 6.7 |
| 10.2 | 20.9 | 16.6 | 33.5 |
| 1.4 | 3.2 | 2.0 | 5.0 |
| 9.4 | 19.0 | 14.0 | 20.1 |
| 1.3 | 3.0 | 1.8 | 4.4 |
| 10.8 | 22.3 | 15.7 | 31.9 |
| 2.2 | 4.5 | 2.6 | 6.3 |
| 4.9 | 11.0 | 8.7 | 16.5 |
| 0.5 | 1.2 | 1.0 | 2.3 |
| 0.0 | 0.0 | 0.0 | 0.0 |
| 0.0 | 0.0 | 0.0 | 0.0 |
| 0.0 | 0.0 | 0.0 | 0.0 |
| 0.0 | 0.0 | 0.0 | 0.0 |
| 7.7 | 26.4 | 10.3 | 33.0 |
| 1.4 | 4.5 | 1.6 | 5.8 |
| 5.2 | 15.4 | 8.3 | 21.5 |
| 0.7 | 1.8 | 1.0 | 2.6 |
| 4.5 | 10.0 | 5.8 | 12.9 |
| 0.9 | 1.9 | 1.1 | 2.5 |
| 7.3 | 20.4 | 9.9 | 26.6 |
| 1.7 | 4.0 | 2.0 | 5.9 |
| 5.1 | 16.1 | 7.8 | 20.4 |
| 1.1 | 2.9 | 1.3 | 4.1 |
| 5.2 | 19.4 | 7.3 | 23.3 |
| 0.8 | 2.2 | 0.7 | 2.8 |
| 6.0 | 14.6 | 9.8 | 20.9 |
| 0.8 | 2.1 | 1.2 | 3.2 |

Appendix G.3. Estimates of morphometrics for 61 species of Australian desert lizards (means, in millimeters).

| Species | Snout-vent length | Tail length | Head length | Head width | Head depth | Jaw length |
|---|---|---|---|---|---|---|
| *Caimanops amphiboluroides* | 71.2 | 109.5* | 17.2 | 12.4 | 9.5 | 19.3 |
| *Ctenophorus clayi* | 41.6 | 55.6 | 10.5 | 8.9 | 6.3 | 11.9 |
| *Ctenophorus fordi* | 48.3 | 104.0 | 10.7 | 8.6 | 5.7 | 12.3 |
| *Ctenophorus inermis* | 75.5 | 77.1 | 15.9 | 14.9 | 11.1 | 19.9 |
| *Ctenophorus isolepis* | 51.4 | 104.8 | 12.0 | 9.8 | 6.8 | 14.0 |
| *Ctenophorus reticulatus* | 72.3 | 99.1* | 15.2 | 14.8 | 10.4 | 19.1 |
| *Ctenophorus scutulatus* | 74.8 | 97.0 | 16.0 | 12.8 | 8.8 | 19.0 |
| *Diporiphora winneckei* | 43.8 | 149.0 | 10.8 | 8.0 | 6.7 | 12.4 |
| *Gemmatophora longirostris* | 64.0 | 238.0 | 15.8 | 10.8 | 8.8 | 19.5 |
| *Moloch horridus* | 89.4 | 66.5 | 13.1 | 13.3 | 10.8 | 15.2 |
| *Pogona minor* | 100.4 | 183.0 | 23.9 | 21.1 | 14.8 | 30.5 |
| *Varanus brevicauda* | 95.3 | 84.2* | 15.5 | 10.2 | 8.3 | 18.1 |
| *Varanus caudolineatus* | 101.2 | 127.0* | 17.0 | 11.4 | 8.2 | 20.2 |
| *Varanus eremius* | 125.8 | 223.6 | 23.3 | 12.9 | 10.3 | 25.7 |
| *Varanus gilleni* | 152.3 | 224.0 | 24.5 | 15.7 | 10.9 | 28.3 |
| *Varanus giganteus* | 726.0 | 999.* | 126.4 | 51.7 | 37.7 | 122.7 |
| *Varanus gouldi* | 283.6 | 407.6 | 47.0 | 25.3 | 19.9 | 53.2 |
| *Varanus tristis* | 229.4 | 410.6 | 38.0 | 19.8 | 15.4 | 43.6 |
| *Ctenotus ariadnae* | 48.9 | 79.9 | 9.2 | 5.7 | 4.5 | 10.8 |
| *Ctenotus atlas* | 58.0 | 113.2* | 11.2 | 6.9 | 5.5 | 12.5 |
| *Ctenotus brooksi* | 37.4 | 72.4 | 8.8 | 4.5 | 3.3 | 9.5 |
| *Ctenotus calurus* | 38.9 | 61.3 | 8.0 | 4.7 | 3.6 | 9.5 |
| *Ctenotus colletti* | 38.5 | 67.0 | 8.0 | 4.2 | 3.1 | 9.1 |
| *Ctenotus dux* | 48.9 | 87.8 | 10.3 | 6.4 | 5.5 | 12.1 |
| *Ctenotus grandis* | 77.0 | 115.2 | 13.9 | 9.2 | 8.2 | 16.4 |
| *Ctenotus helenae* | 75.5 | 125.0 | 13.5 | 8.1 | 6.9 | 15.3 |
| *Ctenotus leae* | 50.0 | 98.3 | 10.2 | 5.4 | 4.2 | 11.3 |
| *Ctenotus leonhardii* | 60.7 | 125.0 | 12.5 | 8.1 | 5.5 | 13.6 |
| *Ctenotus pantherinus* | 71.4 | 95.0 | 12.5 | 8.8 | 8.2 | 14.8 |
| *Ctenotus piankai* | 39.7 | 82.5 | 7.9 | 4.4 | 3.5 | 9.3 |
| *Ctenotus quattuordecimlineatus* | 54.0 | 84.6 | 10.8 | 6.6 | 5.8 | 12.3 |
| *Ctenotus schomburgkii* | 40.2 | 72.9 | 8.6 | 4.7 | 3.5 | 9.7 |
| *Cryptoblepharus plagiocephalus* | 37.5 | 48.0 | 8.0 | 4.9 | 3.3 | 9.6 |
| *Egernia depressa* | 84.7 | 32.4* | 16.5 | 14.6 | 11.1 | 18.8 |
| *Egernia kintorei* | 131.0 | 148.0* | 25.7 | 22.8 | 21.7 | 29.9 |
| *Egernia inornata* | 69.3 | 67.9 | 13.4 | 10.1 | 8.4 | 15.6 |
| *Egernia striata* | 86.5 | 98.2 | 17.3 | 12.4 | 10.3 | 19.6 |
| *Lerista bipes* | 46.9 | 33.3 | 4.5 | 2.6 | 2.2 | 5.5 |
| *Lerista desertorum* | 79.1 | 60.8 | 8.6 | 5.4 | 4.4 | 10.3 |
| *Lerista muelleri* | 40.2 | 41.8 | 4.5 | 2.9 | 2.3 | 5.9 |
| *Menetia greyi* | 26.8 | 22.5 | 3.6 | 2.9 | 2.4 | 5.8 |
| *Morethia butleri* | 43.1 | 66.5 | 8.5 | 5.8 | 4.3 | 10.2 |
| *Omolepida branchialis* | 91.3 | 78.0 | 14.3 | 10.1 | 8.7 | 16.7 |
| *Eremiascincus richardsoni* | 69.4 | 103. | 14.1 | 9.5 | 7.7 | 16.3 |
| *Tiliqua multifasciata* | 233.0 | 109.3 | 44.3 | 40.1 | 27.8 | 51.1 |

| FOREFOOT LENGTH | FOREARM LENGTH | HIND-FOOT LENGTH | HIND-LEG LENGTH |
|---|---|---|---|
| 12.0 | 27.5 | 17.5 | 39.0 |
| 7.4 | 18.3 | 13.5 | 29.9 |
| 8.9 | 22.3 | 24.5 | 50.4 |
| 12.4 | 28.7 | 18.5 | 44.9 |
| 9.4 | 21.6 | 23.5 | 48.4 |
| 12.6 | 29.2 | 19.7 | 45.2 |
| 12.2 | 30.5 | 31.1 | 65.8 |
| 7.9 | 19.2 | 15.6 | 32.5 |
| 11.7 | 27.2 | 25.2 | 55.1 |
| 11.6 | 39.0 | 13.0 | 46.4 |
| 16.1 | 40.9 | 24.3 | 58.1 |
| 10.0 | 20.3 | 10.5 | 25.0 |
| 11.9 | 25.9 | 16.4 | 34.9 |
| 15.1 | 32.3 | 21.3 | 50.2 |
| 16.7 | 35.0 | 21.3 | 45.5 |
| 71.0 | 225.8 | 94.0 | 299.5 |
| 40.6 | 87.7 | 48.9 | 125.4 |
| 29.4 | 62.7 | 40.0 | 91.8 |
| 6.6 | 14.4 | 11.3 | 21.9 |
| 7.6 | 17.1 | 14.1 | 27.4 |
| 5.6 | 12.5 | 10.6 | 20.9 |
| 5.1 | 11.2 | 10.8 | 19.9 |
| 5.4 | 11.7 | 9.4 | 18.3 |
| 6.7 | 15.4 | 13.7 | 25.8 |
| 8.7 | 21.0 | 14.7 | 31.3 |
| 8.0 | 18.5 | 14.1 | 29.7 |
| 7.1 | 16.5 | 15.0 | 29.1 |
| 11.2 | 18.2 | 16.3 | 32.8 |
| 8.0 | 19.4 | 14.0 | 29.3 |
| 5.7 | 12.7 | 10.3 | 19.0 |
| 6.9 | 15.6 | 13.5 | 25.9 |
| 5.6 | 12.3 | 10.9 | 21.0 |
| 5.3 | 12.0 | 7.2 | 15.0 |
| 11.2 | 25.8 | 13.9 | 29.4 |
| 18.0 | 38.0 | 23.5 | 47.5 |
| 7.8 | 20.4 | 12.8 | 27.9 |
| 10.8 | 25.2 | 14.9 | 34.0 |
| 0.0 | 0.0 | 3.0 | 8.2 |
| 0.7 | 2.8 | 4.7 | 12.4 |
| 1.2 | 3.3 | 4.5 | 8.6 |
| 2.5 | 6.5 | 5.0 | 9.4 |
| 4.9 | 11.8 | 8.9 | 17.5 |
| 6.1 | 14.0 | 7.3 | 17.4 |
| 7.4 | 19.2 | 11.6 | 26.3 |
| 10.0 | 39.0 | 11.0 | 40.0 |

Appendix G.3. *(continued)*

| Species | Snout-vent length | Tail length | Head length | Head width | Head depth | Jaw length |
|---|---|---|---|---|---|---|
| *Delma fraseri* | 70.2 | 192.9 | 8.8 | 4.7 | 4.0 | 10.9 |
| *Lialis burtonis* | 176.0 | 113.0 | 18.3 | 6.8 | 5.7 | 20.9 |
| *Pygopus nigriceps* | 162.8 | 171.0 | 12.7 | 7.1 | 6.3 | 15.4 |
| *Diplodactylus ciliaris* | 73.6 | 46.2 | 17.3 | 10.9 | 7.7 | 19.4 |
| *Diplodactylus conspicillatus* | 58.7 | 21.2 | 10.0 | 7.3 | 5.3 | 11.5 |
| *Diplodactylus damaeus* | 48.6 | 41.4 | 10.9 | 6.9 | 4.7 | 12.1 |
| *Diplodactylus elderi* | 38.7 | 18.9 | 9.0 | 6.5 | 4.4 | 10.1 |
| *Diplodactylus pulcher* | 52.0 | 36.1* | 9.8 | 6.6 | 4.8 | 10.8 |
| *Diplodactylus stenodactylus* | 52.0 | 43.0 | 11.5 | 8.2 | 5.3 | 13.3 |
| *Diplodactylus strophurus* | 63.4 | 35.4 | 15.0 | 10.0 | 6.1 | 16.4 |
| *Gehyra variegata* | 51.2 | 44.1 | 11.8 | 8.5 | 5.1 | 13.1 |
| *Heteronotia binoei* | 42.0 | 66.0 | 10.7 | 7.1 | 3.7 | 11.6 |
| *Nephrurus laevissimus* | 60.4 | 20.7 | 16.4 | 12.7 | 7.1 | 16.9 |
| *Nephrurus levis* | 67.1 | 31.7 | 18.4 | 15.1 | 7.7 | 19.2 |
| *Nephrurus vertebralis* | 67.6 | 30.2* | 17.9 | 14.7 | 7.5 | 19.0 |
| *Rhynchoedura ornata* | 46.9 | 32.2 | 8.4 | 6.0 | 4.4 | 9.9 |

*Tail lengths estimated from measurements on fresh material, adjusted for preservation shrinkage.

| FOREFOOT LENGTH | FOREARM LENGTH | HIND-FOOT LENGTH | HIND-LEG LENGTH |
|---|---|---|---|
| 0.0 | 0.0 | 0.0 | 2.8 |
| 0.0 | 0.0 | 0.0 | 1.9 |
| 0.0 | 0.0 | 0.0 | 4.3 |
| 7.5 | 25.7 | 9.2 | 32.5 |
| 5.4 | 16.7 | 6.4 | 17.8 |
| 4.6 | 16.2 | 6.7 | 20.5 |
| 3.8 | 13.5 | 5.3 | 15.6 |
| 5.4 | 18.9 | 7.0 | 21.6 |
| 5.0 | 18.5 | 7.5 | 22.5 |
| 6.7 | 24.8 | 8.7 | 28.4 |
| 5.2 | 14.8 | 7.1 | 19.6 |
| 5.3 | 16.2 | 8.1 | 20.9 |
| 7.4 | 23.4 | 9.7 | 29.9 |
| 8.9 | 28.4 | 11.0 | 33.7 |
| 9.0 | 28.4 | 11.1 | 34.3 |
| 4.3 | 14.4 | 6.9 | 17.8 |

# References

Abrams, P. 1975. Limiting similarity and the form of the competition coefficient. *Theor. Pop. Biol.* 8: 356–375.

———. 1976. Niche overlap and environmental variability. *Math. Biosciences* 28: 357–372.

———. 1977. Density-independent mortality and interspecific competition: A test of Pianka's niche overlap hypothesis. *Amer. Natur.* 111: 539–552.

———. 1980. Some comments on measuring niche overlap. *Ecology* 61: 44–49.

———. 1984. Recruitment, lotteries, and coexistence in coral reef fish. *Amer. Natur.* 123: 44–55.

Ananjeva, N. B. 1977. Morphometrical analysis of limb proportions of five sympatric species of desert lizards (Sauria, *Eremias*) in the southern Balkhash lake region. *Proc. Zool. Inst., Acad. Sci. U.S.S.R.* 74: 3–13.

Aspey, W. P., and J. E. Blankenship. 1977. Spiders and snails and statistical tales: Application of multivariate analyses to diverse ethological data. Pp. 75–120 in *Quantitative Methods in the Study of Animal Behavior*, edited by B. A. Hazlett. New York: Academic Press.

Axelrod, D. I. 1950. *Evolution of Desert Vegetation in Western North America.* Carnegie Inst. of Washington, pub. no. 590: 216–306.

Ballinger, R. E. 1971. Comparative demography of two viviparous lizards (*Sceloporus jarrovi* and *Sceloporus poinsetti*) with consideration of the evolutionary ecology of viviparity in lizards. Ph.D. diss., Texas A. & M. University. *Dissertation Abstracts* (1972), vol. 32(8).

———. 1974. Reproduction of the Texas horned lizard, *Phrynosoma cornutum. Herpetologica* 30: 321–327.

———. 1983. Life-History Variations. Chapter 11 (pp. 241–260) in *Lizard Ecology: Studies of a Model Organism*, edited by R. B. Huey, E. R. Pianka, and T. W. Schoener. Cambridge, Mass.: Harvard University Press.

Ballinger, R. E., and D. R. Clark. 1973. Energy content of lizard eggs and the measurement of reproductive effort. *J. Herpetology* 7: 129–132.

Barbault, R., and G. Halffter. 1981. *Ecology of the Chihuahuan Desert.* Organization of some vertebrate communities. Mexico City: Instituto de Ecologia.

Beard, J. S. 1974. Great Victoria Desert. The vegetation of the Great Victoria desert area (with maps). In *Vegetation Survey of Western Australia.* Nedlands: University of Western Australia Press.

———. 1976. The evolution of Australian desert plants. Chapter 3 in *Evolution of Desert Biota*, edited by D. W. Goodall. Austin: University of Texas Press.

Bender, E. A., T. J. Case, and M. E. Gilpin. 1984. Perturbation experiments in community ecology: Theory and practice. *Ecology* 65: 1–13.

Bennett, A. F., and K. A. Nagy. 1977. Energy expenditure in free-ranging lizards. *Ecology* 58: 697–700.

Blair, W. F. 1960. *The Rusty Lizard*. Austin: University of Texas Press.

Blaisdell, J. P. 1958. Seasonal development and yield of native plants on the upper Snake river plains and their relation to certain climatic factors. *U.S. Dept. Agr. Tech. Bull.*, no. 1190.

Bowler, J. M. 1976. Aridity in Australia: Age, origins and expression in Aeolian landforms and sediments. *Earth Sci. Rev.* 12: 279–310.

Bowler, J. M., G. S. Hope, J. N. Jennings, G. Singh, and D. Walker. 1976. Late Quaternary climates of Australia and New Guinea. *Quaternary Research* 6: 359–394.

Brown, J. H. 1975. Geographical ecology of desert rodents. Chapter 13 (pp. 315–341) in *Ecology and Evolution of Communities*, edited by M. L. Cody and J. M. Diamond. Cambridge, Mass.: Harvard University Press.

———. 1981. Two decades of homage to Santa Rosalia: Towards a general theory of diversity. *Amer. Zoologist* 21: 877–888.

Buckley, R. C. 1981a. Central Australian sandridges. *J. Arid Env.* 4: 91–101.

———. 1981b. Parallel dunefield ecosystems: Southern Kalahari and central Australia. *J. Arid Env.* 4: 287–298.

Burbidge, N. T. 1953. The genus *Triodia* R. Br. (Gramineae). *Australian J. Bot.* 1: 121–184.

Case, T. J. 1975. Species number, density compensation, and colonizing ability of lizards on islands in the Gulf of California. *Ecology* 56: 3–18.

———. 1976. Body size differences between populations of the chuckwalla, *Sauromalus obesus*. *Ecology* 57: 313–323.

———. 1979. Character displacement and coevolution in some *Cnemidophorus* lizards. *Fortschr. Zool.* 25: 235–282.

———. 1982. Ecology and evolution of the insular gigantic chuckwallas, *Sauromalus hispidus* and *Sauromalus varius*. Chapter 11 (pp. 184–212) in *Iguanas of the World*, edited by G. M. Burghardt and A. S. Rand. Park Ridge, N.J.: Noyes Publications.

———. 1983a. Niche overlap and the assembly of island lizard communities. *Oikos* 41: 427–433.

———. 1983b. Geographical ecology of sex and size in *Cnemidophorus* lizards. In *Lizard Ecology: Studies on a Model Organism*, edited by R. B. Huey, E. R. Pianka, and T. W. Schoener. Cambridge, Mass.: Harvard University Press.

Case, T. J., and J. M. Diamond, eds. 1985. *Community Ecology*. New York: Harper and Row.

Case, T. J., and R. Sidell. 1983. Pattern and chance in the structure of model and natural communities. *Evolution* 37: 832–849.

Case, T. J., J. Faaborg, and R. Sidell. 1983. The role of body size in the assembly of West Indian bird communities. *Evolution* 37: 1062–1074.

Caswell, H. 1976. Community structure: A neutral model analysis. *Ecol. Monogr.* 46: 327–354.

Charley, J. L. 1972. The role of shrubs in nutrient cycling. In *Wildland Shrubs—Their Biology and Utilization*, edited by O. B. Williams. U.S.D.A. Forest Service, General Tech. Report, INT-1.

Charley, J. L., and S. L. Cowling. 1968. Changes in soil nutrient status resulting from overgrazing and their consequences in plant communities of semi-arid areas. *Proc. Ecol. Soc. Australia* 3: 28–38.

Charnov, E. L. 1976. Optimal foraging: The marginal value theorem. *Theor. Pop. Biol.* 9: 129–136.

Chesson, P. L. 1982. The stabilizing effect of a random environment. *J. Math. Biol.* 15: 1–36.

———. 1983. Coexistence of competitors in a stochastic environment: The storage effect. *Proc. 1st Int. Conf. Theor. Pop. Biol.*, Edmonton, Alberta.

———. 1985. Environmental variation and the coexistence of species. In *Community Ecology*, edited by T. J. Case and J. M. Diamond. New York: Harper and Row.

Chesson, P. L., and T. J. Case. 1985. Nonequilibrium community theories: Chance, variability, history and coexistence. In *Community Ecology*, edited by T. J. Case and J. M. Diamond. New York: Harper and Row.

Chesson, P. L., and R. R. Warner. 1981. Environmental variability promotes coexistence in lottery competitive systems. *Amer. Natur.* 117: 923–943.

Clark, D. R. 1971. The strategy of tail autotomy in the ground skink, *Lygosoma laterale*. *J. Exper. Zool.* 176: 295–302.

Cody, M. L. 1974. *Competition and the structure of bird communities*. Princeton, N.J.: Princeton University Press.

Cogger, H. G. 1974. Thermal relations of the Mallee dragon *Amphibolurus fordi* (Lacertilia: Agamidae). *Australian J. Zool.* 22: 319–339.

———. 1975. *Reptiles and Amphibians of Australia*. Sydney: Reed.

Collette, B. B. 1961. Correlations between ecology and morphology in anoline lizards from Havana, Cuba, and southern Florida. *Bull. Mus. Comp. Zool.*, (Harvard University) 125: 137–162.

Colwell, R. K., and D. J. Futuyma. 1971. On the measurement of niche breadth and overlap. *Ecology* 52: 567–576.

Colwell, R. K., and E. R. Fuentes. 1975. Experimental studies of the niche. *Ann. Rev. Ecol. and Syst.* 6: 281–310.

Colwell, R. K., and D. W. Winkler. 1984. A null model for null models in biogeography. Chapter 20 (pp. 344–359) in *Ecological communities: Conceptual issues and the evidence*, edited by D. R. Strong, D. Simberloff, L. G. Abele, and A. B. Thistle. Princeton, N.J.: Princeton University Press.

Connell, J. H. 1975. Some mechanisms producing structure in natural communities: A model and evidence from field experiments. Chapter 16 (pp. 460–490) in *Ecology and Evolution of Communities*, edited by M. L. Cody and J. M. Diamond. Cambridge, Mass.: Harvard University Press.

Connell, J. H. 1978. Diversity in tropical rain forests and coral reefs. *Science* 199: 1302–1310.

———. 1980. Diversity and the coevolution of competitors, or the ghost of competition past. *Oikos* 35: 131–138.

Connor, E. F., and D. Simberloff. 1979. The assembly of species communities: Chance or competition? *Ecology* 60: 1132-1140.

Cowles, R. B., and C. M. Bogert. 1944. A preliminary study of the thermal requirements of desert reptiles. *Bull. Amer. Mus. Nat. Hist.* 83: 265–296.

Crocker, R. L., and J. G. Wood. 1947. Some historical influences on the development of the South Australian vegetation communities and their bearing on concepts and classification in ecology. *Trans. Roy. Soc. S. Australia* 17: 91–136.

Dial, B. E., and L. C. Fitzpatrick. 1983. Lizard tail autotomy: Function and energetics of post autotomy tail movement in *Scincella lateralis*. *Science* 219: 391–393.

DiCastri, F., and H. A. Mooney, eds. 1973. *Mediterranean Type Ecosystems.* New York: Springer-Verlag.

Dueser, R. D., and H. H. Shugart. 1979. Niche pattern in a forest floor small-mammal fauna. *Ecology* 60: 108–118.

Dunham, A. E. 1980. An experimental study of interspecific competition between the iguanid lizards *Sceloporus merriami* and *Urosaurus ornatus*. *Ecol. Monogr.* 50: 309–330.

Eckhardt, R. C. 1979. The adaptive syndromes of two guilds of insectivorous birds in the Colorado Rocky Mountains. *Ecol. Monogr.* 49: 129–149.

Emlen, J. M. 1966. The role of time and energy in food preference. *Amer. Natur.* 100: 611–617.

———. 1968. Optimal choice in animals. *Amer. Natur.* 102: 385–390.

Estes, R. 1983. The fossil record and early distribution of lizards. Pp. 365–398 in *Advances in Herpetology and Evolutionary Biology: Essays in Honor of Ernest E. Williams*, edited by A. Rhodin and K. Miyata. Museum of Comparative Zoology, Harvard University.

Feinsinger, P. 1976. Organization of a tropical guild of nectarivorous birds. *Ecol. Monogr.* 46: 257–291.

Findley, J. S. 1973. Phenetic packing as a measure of faunal diversity. *Amer. Natur.* 107: 580–584.

———. 1976. The structure of bat communities. *Amer. Natur.* 110: 129–139.

Finlayson, H. H. 1943. *The Red Centre.* Sydney: Angus and Robertson.

Fox, L. R., and P. A. Morrow. 1981. Specialization: Species property or local phenomenon? *Science* 211: 887–893.

Galloway, R. W., and E. M. Kemp. 1981. Late Cainozoic environments in Australia. Chapter 4 (pp. 51–80) in *Ecological Biogeography in Australia*, edited by A. Keast. The Hague: D. W. Junk.

Gatz, A. J. 1979a. Community organization in fishes as indicated by morphological features. *Ecology* 60: 711–718.

———. 1979b. Ecological morphology of freshwater stream fishes. *Tulane Studies in Zool. and Bot.* 21: 91–124.

Gilpin, M. E. 1974. A Liapunov function for competition communities. *J. Theor. Biol.* 44: 35-48.

Gilpin, M. E., and J. M. Diamond. 1984. Are species co-occurrences on islands non-random, and are null hypotheses useful in community ecology? Chapter 17 (pp. 297–315) in *Ecological Communities: Conceptual issues and the evidence*, edited by D. R. Strong, D. Simberloff, L. G. Abele, and A. B. Thistle. Princeton, N.J.: Princeton University Press.

Grant, P. R. 1967. Bill length variability in birds of the Tres Marias islands, Mexico. *Canadian J. Zool.* 45: 805–815.

———. 1971. Variation in tarsus length of birds in island and mainland regions. *Evolution* 25: 599–614.

———. 1972. Convergent and divergent character displacement. *Biol. J. Linnaean Soc.* 4: 39–68.

Grant, P. R., and I. Abbott. 1980. Interspecific competition, island biogeography and null hypotheses. *Evolution* 34: 332–341.

Green, R. H. 1971. A multivariate statistical approach to the Hutchinsonian niche: Bivalve molluscs of central Canada. *Ecology* 52: 543–556.

———. 1974. Multivariate niche analysis with temporally varying environmental factors. *Ecology* 55: 73–83.

Greene, H. W. 1982. Dietary and phenotypic diversity in lizards: Why are some organisms specialized? Pp. 107–128 in *Environmental Adaptation and Evolution*, edited by D. Mossakowski and G. Roths. Stuttgart: Gustav Fischer.

Greene, H. W., and F. M. Jaksić. 1983. Food-niche relationships among sympatric predators: Effects of level of prey identification. *Oikos* 40: 151–154.

Grinnell, J. 1924. Geography and evolution. *Ecology* 5: 225–229.

Haacke, W. D. 1976. The burrowing geckos of southern Africa, 3 (Reptilia: Gekkonidae). *Annals Transvaal Mus.* 30: 29–39.

Haldane, J.B.S., and J. Huxley. 1927. Pp. 175–176 in *Animal Biology*. London: Oxford University Press.

Hanski, I. 1978. Some comments on the measurement of niche metrics. *Ecology* 59: 168–174.

Harvey, P. H., R. K. Colwell, J. W. Silvertown, and R. M. May. 1983. Null models in ecology. *Ann. Rev. Ecol. and Syst.* 14: 189–211.

Hecht, M. K. 1952. Natural selection in the lizard genus *Aristelliger*. *Evolution* 6: 112–124.

Hespenheide, H. A. 1973. Ecological inferences from morphological data. *Ann. Rev. Ecol. and Syst.* 4: 213–229.

Hoddenbach, G. A., and F. B. Turner. 1968. Clutch size of the lizard *Uta stansburiana* in southern Nevada. *Amer. Midland Natur.* 80: 262–265.

Holmes, R. T., R. E. Bonney, and S. W. Pacala. 1979. Guild structure of the Hubbard Brook bird community: A multivariate approach. *Ecology* 60: 512–520.

Horn, H. S. 1966. Measurement of "overlap" in comparative ecological studies. *Amer. Natur.* 100: 419–424.

Horn, H. S., and R. M. May. 1977. Limits to similarity among coexisting competitors. *Nature* 270: 660–661.

Hotton, N. 1955. A survey of adaptive relationships of dentition to diet in the North American Iguanidae. *Amer. Midland Natur.* 53: 88–114.

Huey, R. B. 1979. Parapatry and niche complementarity of Peruvian desert geckos (*Phyllodactyllus*): The ambiguous role of competition. *Oecologia* 38: 249–259.

Huey, R. B., and E. R. Pianka. 1974. Ecological character displacement in a lizard. *Amer. Zool.* 14: 1127–1136.

———. 1977a. Natural selection for juvenile lizards mimicking noxious beetles. *Science* 195: 201–203.

———. 1977b. Patterns of niche overlap among broadly sympatric versus narrowly sympatric Kalahari lizards (Scincidae: *Mabuya*). *Ecology* 58: 119–128.

———. 1977c. Seasonal variation in thermoregulatory behavior and body temperature of diurnal Kalahari lizards. (With an Appendix by J. A. Hoffman.) *Ecology* 58: 1066–1075.

———. 1981. Ecological consequences of foraging mode. *Ecology* 62: 991–999.

———. 1983. Temporal separation of activity and interspecific dietary overlap. In *Lizard Ecology: Studies of a Model Organism*, edited by R. B. Huey, E. R. Pianka, and T. W. Schoener. Cambridge, Mass.: Harvard University Press.

Huey, R. B., and M. Slatkin. 1976. Costs and benefits of lizard thermoregulation. *Quarterly Rev. Biol.* 51: 363–384.

Huey, R. B., E. R. Pianka, and T. W. Schoener, eds. 1983. *Lizard Ecology: Studies on a Model Organism*. Cambridge, Mass.: Harvard University Press.

Huey, R. B., E. R. Pianka, M. E. Egan, and L. W. Coons. 1974. Ecological shifts in sympatry: Kalahari fossorial lizards (*Typhlosaurus*). *Ecology* 55: 304–316.

Hurlbert, S. H. 1978. The measurement of niche overlap and some relatives. *Ecology* 59: 67–77.

Hutchinson, G. E. 1957. Concluding remarks. *Cold Spring Harbor Symp. on Quant. Biol.* 22: 415–427.

———. 1959. Homage to Santa Rosalia, or why are there so many kinds of animals? *Amer. Natur.* 93: 145–159.

———. 1961. The paradox of the plankton. *Amer. Natur.* 95: 137–145.

Inger, R., and R. K. Colwell. 1977. Organization of contiguous communities of amphibians and reptiles in Thailand. *Ecol. Monogr.* 47: 229–253.

Ivlev, V. S. 1961. *Experimental Feeding Ecology of Fishes*. New Haven, Conn.: Yale University Press.

Jacsić, F. M. 1981. Abuse and misuse of the term "guild" in ecological studies. *Oikos* 37: 397–400.

Joern, A. 1979. Feeding patterns in grasshoppers (Orthoptera: Acrididae): Factors influencing diet specialization. *Oecologia* 38: 325–347.

Joern, A., and L. R. Lawlor. 1980. Food and microhabitat utilization by grasshoppers from arid grasslands: Comparisons with neutral models. *Ecology* 61: 591–599.

———. 1981. Guild structure in grasshopper assemblages based on food and microhabitat resources. *Oikos* 37: 93–104.

Karr, J. R., and F. C. James. 1975. Ecomorphological configurations and convergent evolution in species and communities. Pp. 258–291 in *Ecology and Evolution of Communities*, edited by M. L. Cody and J. M. Diamond. Cambridge, Mass.: Harvard University Press.

Lancaster, N. 1979. Evidence for a widespread late Pleistocene humid period in the Kalahari. *Nature* 279: 145–146.

———. 1984. Aridity in southern Africa: Age, origins and expression in landforms and sediments. In *Late Cainozoic Palaeoclimates of the Southern Hemisphere*, edited by J. C. Vogel. Rotterdam: Ballcema.

Lawlor, L. R. 1979. Direct and indirect effects of *n*-species competition. *Oecologia* 43: 355–364.

———. 1980a. Overlap, similarity, and competition coefficients. *Ecology* 61: 245–251.

———. 1980b. Structure and stability in natural and randomly-constructed competitive communities. *Amer. Natur.* 116: 394–408.

Lein, M. R. 1972. A trophic comparison of avifaunas. *Syst. Zool.* 21: 415–427.

Leistner, O. A. 1967. The plant ecology of the southern Kalahari. *Bot. Survey S. Africa, Memoirs* 38: 1–172.

Levine, S. H. 1976. Competitive interactions in ecosystems. *Amer. Natur.* 110: 903–910.

Levins, R. 1968. *Evolution in Changing Environments*. Princeton, N.J.: Princeton University Press.

Licht, P., W. R. Dawson, V. H. Shoemaker, and A. R. Main. 1966. Observations on the thermal relations of Western Australian lizards. *Copeia* 1966: 97–110.

Linton, L. R., R. W. Davies, and F. J. Wrona. 1981. Resource utilization indices: An assessment. *J. Animal Ecol.* 50: 283–292.

Lowe, C. H. 1968. Fauna of desert environments. Pp. 567–645 in *Deserts of the World: An Appraisal of Research into their Physical and Biological Environments*, edited by W. G. McGinnies, B. J. Goldman, and P. Paylore. Tucson: University of Arizona Press.

Lundelius, E. L. 1957. Skeletal adaptations in two species of *Sceloporus*. *Evolution* 11: 65–83.

MacArthur, R. H. 1965. Patterns of species diversity. *Biol. Rev.* 40: 510–533.

———. 1970. Species packing and competitive equilibrium for many species. *Theor. Pop. Biol.* 1: 1–11.

MacArthur, R. H. 1972. *Geographical Ecology*. New York: Harper and Row.

MacArthur, R. H., and R. Levins. 1967. The limiting similarity, convergence and divergence of coexisting species. *Amer. Natur.* 101: 377–385.

MacArthur, R. H., and E. R. Pianka. 1966. On optimal use of a patchy environment. *Amer. Natur.* 100: 603–609.

MacArthur, R. H., and E. O. Wilson. 1967. *The Theory of Island Biogeography*. Princeton, N.J.: Princeton University Press.

Marion, K. R., and O. J. Sexton. 1971. The reproductive cycle of the lizard *Sceloporus malachiticus* in Costa Rica. *Copeia* 1971: 517–526.

Maury, M. E. 1981. Food partition of lizard communities at the Bolson de Mapimi (Mexico). Pp. 119–142 in *Ecology of the Chihuahuan Desert*, edited by R. Barbault and G. Halffter. Mexico City: Instituto de Ecologia.

Maury, M. E., and R. Barbault. 1981. The spatial organization of the lizard community of the Bolson de Mapimi (Mexico). Pp. 79–87 in *Ecology of the Chihuahuan Desert*, edited by R. Barbault and G. Halffter. Mexico City: Instituto de Ecologia.

May, R. M. 1974. On the theory of niche overlap. *Theor. Pop. Biol.* 5: 297–332.

———. 1975. Some notes on estimating the competition matrix, α. *Ecology* 56: 737–741.

May, R. M., and R. H. MacArthur. 1972. Niche overlap as a function of environmental variability. *Proc. Nat. Acad. Sci. U.S.A.* 69: 1109–1113.

Mayhew, W. W. 1967. Comparative reproduction in three species of the genus *Uma*. Pp. 45–65 in *Lizard Ecology: A Symposium*, edited by W. W. Milstead. Columbia: University of Missouri Press.

———. 1968. Biology of desert amphibians and reptiles. Pp. 195–356 in *Desert Biology*, vol. 1, edited by G. W. Brown. New York: Academic Press.

Mellado, J., F. Amores, F. F. Parreno, and F. Hiraldo. 1975. The structure of a Mediterranean lizard community. *Donana Acta Vert.* 2: 145–160.

Milewski, A. V. 1981. A comparison of reptile communities in relation to soil fertility in the Mediterranean and adjacent arid parts of Australia and southern Africa. *J. Biogeography* 8: 493–503.

Mosauer, W. 1932. Adaptive convergence in the sand reptiles of the Sahara and of California. *Copeia* 1932: 72–78.

Norris, K. S., and J. L. Kavanau. 1966. The burrowing of the western shovel-nosed snake, *Chionactis occipitalis*, and the undersand environment. *Copeia* 1966: 650–664.

Orians, G. H. 1980. Micro and macro in ecological theory. *BioScience* 30: 79.

Orians, G. H., and H. S. Horn. 1969. Overlap in foods and foraging of four species of blackbirds in the Potholes of central Washington. *Ecology* 50: 930–938.

Orians, G. H., and O. T. Solbrig, eds. 1977. *Convergent Evolution in Warm Deserts: An Examination of Strategies and Patterns in Deserts of Argentina and the United States*. U.S./I.B.P. Synthesis Series, vol. 3. Stroudsburg, Pa.: Dowden, Hutchinson and Ross, Inc.

Parker, W. S., and E. R. Pianka. 1973. Notes on the ecology of the iguanid lizard, *Sceloporus magister*. *Herpetologica* 29: 143–152.

Parker, W. S., and E. R. Pianka. 1974. Further ecological observations on the western banded gecko, *Coleonyx variegatus*. *Copeia* 1974: 528–531.

———. 1975. Comparative ecology of populations of the lizard *Uta stansburiana*. *Copeia* 1975: 615–632.

———. 1976. Ecological observations on the leopard lizard *Crotaphytus wislizeni* in different parts of its range. *Herpetologica* 32: 95–114.

Pearson, L. C. 1965. Primary production in grazed and ungrazed desert communities of eastern Idaho. *Ecology* 45: 278–286.

Pearson, O. P., and C. P. Ralph. 1978. The diversity and abundance of vertebrates along an altitudinal gradient in Peru. *Memorias del Museo de Historia Natural "Javier Prado,"* no. 18: 1–97.

Pianka, E. R. 1966. Convexity, desert lizards, and spatial heterogeneity. *Ecology* 47: 1055–1059.

———. 1967. On lizard species diversity: North American flatland deserts. *Ecology* 48: 333–351.

———. 1968. Notes on the biology of *Varanus eremius*. *W. Australian Natur.* 11: 39–44.

———. 1969a. Notes on the biology of *Varanus caudolineatus* and *Varanus gilleni*. *W. Australian Natur.* 11: 76–82.

———. 1969b. Habitat specificity, speciation, and species density in Australian desert lizards. *Ecology* 50: 498–502.

———. 1969c. Sympatry of desert lizards (*Ctenotus*) in Western Australia. *Ecology* 50: 1012–1030.

———. 1970a. Notes on *Varanus brevicauda*. *W. Australian Natur.* 11: 113–116.

———. 1970b. Notes on the biology of *Varanus gouldi flavirufus*. *W. Australian Natur.* 11: 141–144.

———. 1970c. Comparative autecology of the lizard *Cnemidophorus tigris* in different parts of its geographic range. *Ecology* 51: 703–720.

———. 1971a. Comparative ecology of two lizards. *Copeia* 1971: 129–138.

———. 1971b. Notes on the biology of *Varanus tristis*. *W. Australian Natur.* 11: 180–183.

———. 1971c. Ecology of the agamid lizard *Amphibolurus isolepis* in Western Australia. *Copeia* 1971: 527–536.

———. 1971d. Notes on the biology of *Amphibolurus cristatus* and *Amphibolurus scutulatus*. *W. Australian Natur.* 12: 36–41.

———. 1971e. Lizard species density in the Kalahari desert. *Ecology* 52: 1024–1029.

———. 1972. Zoogeography and speciation of Australian desert lizards: An ecological perspective. *Copeia* 1972: 127–145.

Pianka, E. R. 1973. The structure of lizard communities. *Ann. Rev. Ecol. and Syst.* 4: 53–74.

———. 1974. Niche overlap and diffuse competition, *Proc. Nat. Acad. Sci. U.S.A.* 71: 2141–2145.

———. 1975. Niche relations of desert lizards. Chapter 12 (pp. 292–314) in *Ecology and Evolution of Communities*, edited by M. Cody and J. Diamond. Cambridge, Mass.: Harvard University Press.

———. 1976a. Competition and niche theory. Chapter 7 (pp. 114–141) in *Theoretical Ecology: Principles and Applications*, edited by R. M. May. London: Blackwell.

———. 1976b. Natural selection of optimal reproductive tactics. *Amer. Zool.* 16: 775–784.

———. 1977. Reptilian species diversity. Chapter 1 (pp. 1–34) in *Biology of the Reptilia*, edited by C. Gans and D. W. Tinkle. New York: Academic Press.

———. 1979. Diversity and niche structure in desert communities. Chapter 10 (pp. 321–341) in *Arid-Land Ecosystems: Structure, Function and Management*, edited by R. Perry and D. Goodall. London: Cambridge University Press.

———. 1980. Guild structure in desert lizards. *Oikos* 35: 194–201.

———. 1981. Diversity and adaptive radiations of Australian desert lizards. Chapter 50 (pp. 1375–1392) in *Ecological Biogeography in Australia*, edited by A. Keast. The Hague: D. W. Junk.

———. 1982. Observations on the ecology of *Varanus* in the Great Victoria desert. *W. Australian Natur.* 15: 37–44.

———. 1985. Some intercontinental comparisons of desert lizards. *National Geographic Research* 1(4): 490–504.

Pianka, E. R., and W. F. Giles. 1982. Notes on the biology of two species of nocturnal skinks, *Egernia inornata* and *Egernia striata*, in the Great Victoria desert. *W. Australian Natur.* 15: 44–49.

Pianka, E. R., and R. B. Huey. 1971. Bird species density in the Kalahari and the Australian deserts. *Koedoe* 14: 123–130.

———. 1978. Comparative ecology, niche segregation, and resource utilization among gekkonid lizards in the southern Kalahari. *Copeia* 1978: 691–701.

Pianka, E. R., and W. S. Parker. 1972. Ecology of the iguanid lizard *Callisaurus draconoides. Copeia* 1972: 493–508.

———. 1975. Ecology of horned lizards: A review with special reference to *Phrynosoma platyrhinos. Copeia* 1975: 141–162.

———. 1975. Age-specific reproductive tactics. *Amer. Natur.* 109: 453–464.

Pianka, E. R., and H. D. Pianka. 1970. The ecology of *Moloch horridus* (Lacertilia: Agamidae) in Western Australia. *Copeia* 1970: 90–103.

———. 1976. Comparative ecology of twelve species of nocturnal lizards (Gekkonidae) in the Western Australian desert. *Copeia* 1976: 125–142.

Pianka, E. R., and J. J. Schall. 1981. Species densities of terrestrial vertebrates in Australia. In *Ecological Biogeography in Australia*, edited by A. Keast. The Hague: D. W. Junk.

Pianka, E. R., R. B. Huey, and L. R. Lawlor. 1979. Niche segregation in desert lizards. Chapter 4 (pp. 67–115) in *Analysis of Ecological Systems*, edited by D. J. Horn, R. Mitchell, and G. R. Stairs. Columbus: Ohio State University Press.

Pianka, H. D., and E. R. Pianka. 1970. Bird censuses from desert localities in Western Australia. *Emu* 70: 17–22.

Pielou, E. C. 1969. *An Introduction to Mathematical Ecology*. New York: Wiley-Interscience.

———. 1972. Niche width and niche overlap: A method for measuring them. *Ecology* 53: 687–692.

———. 1976. *Mathematical Ecology*. New York: Wiley.

———. 1984. *The Interpretation of Ecological Data: A Primer on Classification and Ordination*. New York: Wiley.

Pimm, S. L. 1982. *Food Webs*. London: Chapman and Hall.

———. 1984. The complexity and stability of ecosystems. *Nature* 307: 321–326.

Polis, G. A., W. D. Sissom, and S. J. McCormick. 1981. Predators of scorpions: Field data and a review. *J. Arid Env.* 4: 309–326.

Porter, W. P., J. W. Mitchell, W. A. Beckman, and C. B. DeWitt. 1973. Behavioral implications of mechanistic ecology. *Oecologia* 13: 1–54.

Pough, F. H. 1980. The advantages of ectothermy for tetrapods. *Amer. Natur.* 115: 92–112.

Quinn, J. F., and A. E. Dunham. 1983. On hypothesis testing in ecology and evolution. *Amer. Natur.* 122: 602-617.

Rand, A. S. 1967. Predator-prey interactions and the evolution of aspect diversity. *Atlas Simp. Sobre Biota Amazonica* 5: 73–83.

Rappoldt, C., and P. Hogeweg. 1980. Niche packing and number of species. *Amer. Natur.* 116: 480–492.

Recher, H. F. 1969. Bird species diversity and habitat diversity in Australia and North America. *Amer. Natur.* 103: 75–80.

Regal, P. 1978. Behavioral differences between reptiles and mammals: An analysis of activity and mental capabilities. Pp. 183–202 in *Behavior and Neurology of Lizards*, edited by N. Greenberg and P. MacLean. Nat. Inst. Mental Health, U.S. Dept. H.E.W.

Ricklefs, R. E., and K. O'Rouke. 1975. Aspect diversity in moths: A temperate-tropical comparison. *Evolution* 29: 313–324.

Ricklefs, R. E., and J. Travis. 1980. A morphological approach to the study of avian community organization. *Auk* 97: 321–338.

Ricklefs, R. E., D. Cochran, and E. R. Pianka. 1981. A morphological analysis of the structure of communities of lizards in desert habitats. *Ecology* 62: 1474–1483.

Robinson, M. D. 1973. Nutritional ecology of a Namib desert sand dune lizard, *Aporosaura anchietae*. Abstract, Second International Congress of Ecology (INTECOL), Jerusalem.

Robinson, M. D., and A. B. Cunningham. 1978. Comparative diet of two Namib desert sand lizards (Lacertidae). *Madoqua* 11: 41–53.

Root, R. B. 1967. The niche exploitation pattern of the Blue-gray Gnatcatcher. *Ecol. Monogr.* 37: 317–350.

Rosenberg, H. I., and A. P. Russell. 1980. Structural and functional aspects of tail squirting: A unique defense mechanism of *Diplodactylus* (Reptilia: Gekkonidae). *Can. J. Zool.* 58: 865–881.

Rosenzweig, M. L. 1968. Net primary productivity of terrestrial communities: Prediction from climatological data. *Amer. Natur.* 102: 683–718.

Roughgarden, J. 1972. Evolution of niche width. *Amer. Natur.* 106: 683–718.

———. 1974. Niche width: Biogeographic patterns among *Anolis* lizard populations. *Amer. Natur.* 108: 429–442.

———. 1976. Resource partitioning among competing species—a coevolutionary approach. *Theor. Pop. Biol.* 9: 388–424.

Sage, R. D. 1972. The origin and structure of the desert ecosystem: The lizard component. Pp. 149–178 in *Origin and Structure of Ecosystems*, Technical Report 72–6, Papers presented at the Annual Meeting of the Structure of Ecosystems Program, held March 1972, Tucson, Arizona.

———. 1974. The structure of lizard faunas: Comparative biologies of lizards in two Argentina deserts. Ph.D. Diss., University of Texas at Austin.

Sale, P. F. 1974. Overlap in resource use, and interspecific competition. *Oecologia* 17: 245–256.

Schall, J. J. 1977. Thermal ecology of five sympatric whiptail lizards *Cnemidophorus* (Sauria: Teiidae). *Herpetologica* 33: 261–272.

———. 1978. Reproductive strategies in sympatric whiptail lizards (*Cnemidophorus*): Two parthenogenetic and three bisexual species. *Copeia* 1978: 108–116.

Schall, J. J., and E. R. Pianka. 1978. Geographical trends in numbers of species. *Science* 201: 679–686.

———. 1980. Evolution of escape behavior diversity. *Amer. Natur.* 115: 551–566.

Schoener, T. W. 1965. The evolution of bill size differences among sympatric congeneric species of birds. *Evolution* 19: 189–213.

———. 1968. The *Anolis* lizards of Bimini: Resource partitioning in a complex fauna. *Ecology* 49: 704–726.

———. 1970. Non-synchronous spatial overlap of lizards in patchy habitats. *Ecology* 51: 408–418.

———. 1971. Theory of feeding strategies. *Ann. Rev. Ecol. and Syst.* 2: 369–404.

———. 1974a. Resource partitioning in ecological communities. *Science* 185: 27–39.

———. 1974b. Some methods for calculating competition coefficients from resource utilization spectra. *Amer. Natur.* 108: 332–340.

———. 1974c. The compression hypothesis and temporal resource partitioning. *Proc. Nat. Acad. Sci. U.S.A.* 71: 4169–4172.

———. 1975. Presence and absence of habitat shift in some widespread lizard species. *Ecol. Monogr.* 45: 232–258.

———. 1977. Competition and the niche. Chapter 2 (pp. 35–136) in *Biology of the Reptilia*, vol. 7, edited by D. W. Tinkle and C. Gans. New York: Academic Press.

———. 1979. Inferring the properties of predation and other injury-producing agents from injury frequencies. *Ecology* 60: 1110–1115.

———. 1982. The controversy over interspecific competition. *Amer. Sci.* 70: 586–595.

———. 1983. Field experiments on interspecific competition. *Amer. Natur.* 122: 240–285.

———. 1984. Size differences among sympatric, bird-eating hawks: A worldwide survey. In *Ecological Communities: Conceptual Issues and the Evidence*, edited by D. R. Strong, D. Simberloff, L. G. Abele and A. B. Thistle. Princeton, N.J.: Princeton University Press.

Schoener, T. W., R. B. Huey, and E. R. Pianka. 1979. A biogeographic extension of the compression hypothesis: Competitors in narrow sympatry. *Amer. Natur.* 113: 295–298.

Shannon, C. E. 1949. The mathematical theory of communication. In *The Mathematical Theory of Communication*, edited by C. E. Shannon and W. Weaver. Urbana: University of Illinois Press.

Shreve, F. 1942. The desert vegetation of North America. *Bot. Rev.* 8: 195–244.

———. 1951. *Vegetation of the Sonoran Desert.* Carnegie Inst. of Washington, pub. no. 590. (Reprinted in 1964 by Stanford University Press, Palo Alto, Calif.)

Simberloff, D. S., and W. Boecklen. 1981. Santa Rosalia reconsidered. *Evolution* 35: 1206–1228.

Simpson, E. H. 1949. Measurement of diversity. *Nature* 163: 688.

Snyder. R. C. 1952. Quadrupal and bipedal locomotion in lizards. *Copeia* 1952: 64–70.

Soule, M., and B. R. Stewart. 1970. The "niche-variation" hypothesis: A test and alternatives. *Amer. Natur.* 104: 85–97.

Storr, G. M., L. A. Smith, and R. E. Johnstone. 1981. *Lizards of Western Australia*, vol. I: *Skinks*. Nedlands: University of Western Australia Press.

———. 1983. *Lizards of Western Australia*, vol. II: *Dragons and Monitors.* Nedlands: University of Western Australia Press.

———. 1985. *Lizards of Western Australia*, vol. III: *Geckos and Pygopodids.* Nedlands: University of Western Australia Press.

Strong, D. R., L. A. Szyska, and D. S. Simberloff. 1979. Tests of community-wide character displacement against null hypotheses. *Evolution* 33: 897-913.

Strong, D. R., D. Simberloff, L. G. Abele, and A. B. Thistle, ed. 1984. *Ecological Communities: Conceptual Issues and the Evidence.* Princeton, N.J.: Princeton University Press.

Sugihara, G. 1980. Minimal community structure: An explanation of species abundance patterns. *Amer. Natur.* 116: 770–787.

———. 1982. Holes in niche space: A derived assembly rule and its relation to intervality. Pp. 25–35 in *Current Trends in Food Web Theory*, edited by D. L. De Angelis, P. M. Post, and G. Sugihara. Oak Ridge National Laboratory, publication 5983. National Technical Information Service, U.S. Dept. of Commerce.

———. 1984. Graph theory, homology and food webs. *Proc. Symposia in Applied Math.* 30: 83–101 (American Mathematical Society).

Taylor, W. D. 1979. Sampling data on the bactivorous ciliates of a small pond compared to neutral models of community structure. *Ecology* 60: 876–883.

Tinkle, D. W. 1967. *The Life and Demography of the Side-Blotched Lizard, Uta stansburiana.* Misc. Publ. Mus. Zool. (Univ. of Michigan), no. 132.

———. 1969. The concept of reproductive effort and its relation to the evolution of life histories of lizards. *Amer. Natur.* 103: 501–516.

Tinkle, D. W., and N. F. Hadley. 1975. Lizard reproductive effort: Caloric estimates and comments on its evolution. *Ecology* 56: 427–434.

Tinkle, D. W., H. M. Wilbur, and S. G. Tilley. 1970. Evolutionary strategies in lizard reproduction. *Evolution* 24: 55–74.

Toft, C. A. 1985. Resource partitioning in amphibians and reptiles. *Copeia* 1985: 1-21.

Tramer, E. J. 1969. Bird species diversity: Components of Shannon's formula. *Ecology* 50: 927–929.

Vandermeer, J. H. 1972. Niche theory. *Ann. Rev. Ecol. and Syst.* 3: 107–132.

Van Valen, L. 1965. Morphological variation and the width of the ecological niche. *Amer. Natur.* 100: 377–389.

Van Valen, L., and P. R. Grant. 1970. Variation and niche width reexamined. *Amer. Natur.* 104: 589–590.

Vitt, L. J. 1977. Caloric content of lizard and snake (Reptilia) eggs and bodies and the conversion of weight to caloric data. *J. Herpetology* 12: 65–72.

Vitt, L. J., and J. D. Congdon. 1978. Body shape, reproductive effort, and relative clutch mass in lizards: Resolution of a paradox. *Amer. Natur.* 112: 595–608.

Vitt, L. J., and H. J. Price. 1982. Ecological and evolutionary determinants of relative clutch mass in lizards. *Herpetologica* 38: 237–255.

Vitt, L. J., J. D. Congdon, and N. Dickson. 1977. Adaptive strategies and energetics of tail autotomy in lizards. *Ecology* 58: 326–337.

Walter, H. 1939. Grassland, Savanne und Busch der arideren Teile Afrikas in ihrer ökologischen Bedingtheit. *Jahrbücher für wiss. Bot.* 87: 750–860.

———. 1955. Le facteur eau dans les régiones arides et sa signification pour l'organisation de la végétation dans les contrées sub-tropicales. Pp. 27–39 in *Colloques Int. du Centre National de la Recherche Sci.*, vol. 59, Div. Ecol. du Monde, Centre Nat. de la Recherche Sci., Paris.

———. 1964. Productivity of vegetation in arid countries, the savanna problem, and bush encroachment after overgrazing. *Proc. Papers I.U.C.N. 9th Tech. Meeting.* I.U.C.N. Tech. Publ., new series no. 4: 221–229.

Ward, J. D., M. K. Seely, and N. Lancaster. 1983. On the antiquity of the Namib. *S. African J. Sci.* 79: 175–183.

Werner, Y. 1969. Eye size in geckos of various ecological types (Reptilia: Gekkonidae and Sphaerodactylidae). *Israel J. Zool.* 18: 291–316.

———. 1977. Ecological comments on some gekkonid lizards of the Namib desert, South West Africa. *Madoqua* 10: 157–169.

Wiens, J. A. 1977. On competition and variable environments. *Amer. Sci.* 65: 590–597.

———. 1983. Avian community ecology: An iconoclastic view. Chapter 10 in *Perspectives in Ornithology*, edited by A. H. Brush and G. A. Clark. Cambridge, Eng., and New York: Cambridge University Press.

Williams, G. E. 1973. Late Quaternary piedmont sedimentation, soil formation and paleoclimates in arid South Australia. *Z. Geomorph. N.F.* 17: 102–125.

Willson, M. F. 1969. Avian niche size and morphological variation. *Amer. Natur.* 103: 531–542.

Winkworth, R. E. 1967. The composition of several arid spinifex grasslands of central Australia in relation to rainfall, soil water relations, and nutrients. *Australian J. Bot.* 15: 107–130.

Yoshiyama, R. M., and J. Roughgarden. 1977. Species packing in two dimensions. *Amer. Natur.* 111: 107–121.

Zweifel, R. G., and C. H. Lowe. 1966. The ecology of a population of *Xantusia vigilis*, the desert night lizard. *Amer. Mus. Novitiates* 2247: 1–57.

# Author Index

# Subject Index

*Library of Congress Cataloging-in-Publication Data*

Pianka, Eric R.
Ecology and natural history of desert lizards.

Bibliography: p.
Includes indexes.
1. Lizards—Ecology. 2. Desert fauna—Ecology.
3. Niche (Ecology) 4. Biotic communities.
5. Reptiles—Ecology. I. Title.
QL666.L2P53 1986 597.95'0452652 85–19097
ISBN 0–691–08148–4 (alk. paper)
ISBN 0–691–08406–8 (pbk.)